Contents

Introduction to Advanced Chemistry 20
    Properties of matter 20
        Physical properties 20
        Chemical properties 20
    Basic definitions 20
    Fundamental forces 20
    Energy changes 21
Chemical calculations (introduction and review) 21
    Units 21
        Mass 21
        Volume 21
1.1   Introduction to the particulate nature of matter and chemical change 22
    Syllabus 22
Matter 22
States of matter 23
    Kinetic Theory of Matter 23
    The kelvin scale for temperature 24
    State symbols 25
    Pure substance - Elements 25
    Pure substance - Compounds 25
Mixtures 27
    Homogeneous mixtures 27
    Heterogeneous mixtures 27
Ions 27
Cations and their charges 28
Anions and their charges 28
    Formulae of ionic compounds 29
    Common acids 29
    Balancing chemical equations 29
    Ionic equations 31

Solubility of ionic compounds 32
Atom Economy 32
1.2 The mole concept 33
    Syllabus 33
    The mole and the Avogadro constant 34
    Mole ratios 34
    Measurements and units 35
        Systeme International (SI) units 35
        Prefixes for units 35
    Important conversions 36
    Important constants 36
        Avogadro's number (N_A) 36
    Ratio of elements 36
The mole 37
Formulas 37
    Molar mass 37
    Relative atomic mass (A_r) 38
    Relative molecular mass (M_r) 39
    Examples of relative molecular mass: 39
    Relative formula mass 39
    Examples of relative formula mass: 39
    Calculating number of moles from mass 40
    Calculating the mass of substance 40
    Percentage composition 41
    Empirical and molecular formulae 41
        Calculating empirical and molecular formulae 42
1.3   Reacting masses and volumes 45
    Syllabus 45
Calculating reacting masses 46
    Using masses of reactants to work out the balanced equation for a reaction. 48
Percentage yield 49
Calculations involving Gas Volumes 50

- Gas Volumes 50
- The gas laws 52
  - Boyle's law 52
  - Charles' law 52
  - Gay-Lussac (Pressure) law 52
  - Combined gas law 53
  - Ideal gas equation 53
- Concentrations of Solutions 54
  - Very small concentrations 55
  - Titration calculations 56
- 2.1 The nuclear atom 57
  - Syllabus 57
- The basic structure of an atom 57
- Timeline of discovery of sub-atomic particles 58
- Models of the atom 58
  - Thomson's 'Plum Pudding' model (1807) 58
  - Dalton's Solid Sphere model (1808) 59
  - Rutherford's Planetary model (1911) 60
    - Rutherford's gold foil experiment 60
- Size of atoms 61
- Masses (definitions) 61
  - Relative atomic mass (A_r) 61
  - Relative molecular mass (M_r) 61
  - Relative isotopic mass 61
- Isotopes 61
  - The nuclear symbol (isotopic notation) 62
  - Naming isotopes 62
- Radioisotopes 62
  - Uses of radioisotopes 62
- Mass Spectrometer 63
  - Sample applications 64
  - Other uses of a Mass Spectrometer 66

2.2 Electronic configuration 67
Electromagnetic spectrum 67
    Different types of electromagnetic spectra 68
Electron arrangement 68
    Energy in discrete parcels (quantization) – Max Planck 68
    Energy Levels and the Line Spectra (of hydrogen) – Niels Bohr 69
    Emission spectra 69
Ionisation energy (IE or ΔH_i) 72
Sublevels and orbitals 73
    n=1 73
    n=2 73
    n=3 73
    n=4 73
Heisenberg's Uncertainty Principle 74
Schrödinger's equation 74
Electron spin and the spin magnetic quantum number, m_s 74
Order of energies for the orbitals 75
Filling orbitals (the aufbau method) 76
    Unabbreviated electron configurations 78
    Abbreviated electron configurations 78
    More examples 78
    Validity of electronic configurations 78
12.1 Electrons in atoms 79
    Syllabus 79
Wave-particle duality (de Broglie equation) 79
Heisenberg's uncertainty principle (extension) 80
Ionisation energy (IE) 80
    Successive ionisation energy 81
3.1 Periodic Table 82
    Historical development – Döbereiner (Law of Triads) 83
Development of the Periodic Table 83
    Historical development – Newlands (Law of Octaves) 83

Historical development – Mendeleev (Periodic Law) 84

Historical development – Moseley's modifications 84

Metals 84

    Physical properties 84

    Chemical properties 84

Non-metals 85

Metalloids 85

Valence electrons (outer-shell electrons) 85

The Modern Periodic Table 85

Physical properties 85

    Electronic configurations and valency 86

    Metallic properties 86

3.2 Periodic trends 87

    Syllabus 87

    Elemental structure 87

Forces on an electron in an atom 88

    Shielding effect and effective nuclear charge 88

        Across Period 1 89

        Across Period 2 89

        Down a group 89

    Atomic and Ionic Radii 90

        Bonding atomic radius (covalent radius), R_b 90

        Non-bonding atomic radius (van der Waals' radius), R_{nb} 90

        Atomic Radii trends 91

        Ionic Radii trends 92

Melting and boiling point trends 93

    Across periods 2 and 3 93

    Down Groups 1 and 2 (metals) 94

    Down Group 17 94

Electronegativity 95

First ionisation energy trends 96

    Down a group 96

- Across a period 97
  - First Ionisation Energies and reactivity 97
- Electron affinity (E_{ea}) 97
- Metallic properties 98
- Chemical properties 99
  - Group 0: the noble gases 99
  - Group 1: the alkali metals 99
  - Group 17: the halogens 101
  - Tests for halides 103
- Period 3 Oxides 104
  - Formulae 104
  - Structures 104
  - Acid-base character 105
- 13.1 First-row d-block elements 107
  - Syllabus 107
- Transition metals 108
  - Electronic configuration of atoms 109
  - Electronic configuration of ions 111
    - Evidence for electronic congifurations 112
- Oxidation states 112
  - Variable oxidation states 113
  - Most common oxidation states 113
- General physical properties 114
- General chemical properties 114
- Complexes and Ligands 115
  - Coordination number 116
- Charge on the central ion 116
  - Shapes of Complexes 116
  - Types of ligands 118
    - Uses of EDTA 119
  - Common complexes 119
- Catalytic activity of transition metals 120

Heterogeneous catalysts 120

    Advantages of heterogeneous catalysts 120

Homogeneous catalysts 121

    Paramagnetic properties 122

    Diamagnetic properties 122

13.2 Coloured compounds 123

    Syllabus 123

    Evolving theories on the formation of transition metal complexes 124

    Reasons for coloured compounds 124

    Crystal Field Theory 125

    Octahedral field Splitting Pattern 126

    Spectrochemical series 126

    Crystal Field Theory (CFT) – Electronic configuration 127

        M^{2+} complexes 128

        M^{3+} complexes 128

        Crystal Field Theory (CFT) – Coloured complexes (d-to-d electronic transition)
.... 129

4.1 Ionic bonding and structure 130

    Syllabus 130

    Formation of ions 130

        Oxidation is Loss (OIL) 131

        Reduction is Gain (RIG) 131

        Giant ionic lattice 132

    Cations and their charges 133

    Anions and their charges 133

    Strength of ionic bonding (melting points of ionic compounds) 134

4.2 Covalent bonding 135

    Syllabus 135

    Formation of covalent bonds 135

    Electronegativity 136

    Lewis structures 137

    Dative covalent (co-ordinate) bonding 140

Coordinate (dative covalent) bonding – Al_2Cl_6 dimer 141
Covalent bond strength 141
Bond polarity due to difference in electronegativity 142
    Dipole notation 143
Microwaves in cooking 143
Polarity in molecules 144
    Symmetrical molecules – non-polar 144
    Asymmetrical molecules – polar 144

4.3 Covalent structures 145
Syllabus 145
Shapes of molecules 146
(Valence shell) Electron-pair repulsion theory (VSEPR) 146
Molecules with 3 electron domains 147
Molecules with 4 electron domains 148
Delocalization of electrons and resonance structures 150
    Allotropes of Carbon – Covalent Network Solids (Graphene) 156
    Covalent Network solid – silicon dioxide/silica (quartz) 157

4.4 Intermolecular forces 158
Syllabus 158
Summary of interactions between atoms, molecules and ions 158
London (dispersion) forces 159
    London (dispersion) forces - Factors affecting 160
Dipole-dipole attraction 162
Hydrogen bonding 163
    Hydrogen bonding through fluorine 163
    Hydrogen bonding through oxygen 164
    Hydrogen bonding through nitrogen 165
Alcohols 167

4.5 Metallic bonding 168
Syllabus 168
Drude-Lorentz model 169
Physical properties 169

Malleability and Ductility ... 169

Alloys ... 170

Electrical conductivity ... 170

Trends in melting points ... 170

14.1 Further aspects of covalent bonding and structure ... 171

Syllabus ... 171

Formal Charge ... 172

Formal Charge – incomplete octets (fewer than 8 electrons) ... 173

Formal Charge – expanded octets (more than 8 electrons) ... 174

Molecular orbital theory (MOT) – Bonding orbitals ... 177

Sigma (σ)-bonding ... 177

Pi (π)-bonding ... 178

Molecular orbital theory (MOT) – Antibonding orbitals ... 178

Molecular orbital theory (MOT) – Non-bonding situations ... 179

Evidence of delocalization of electrons and resonance structures (bond order) ... 180

14.2 Hybridization ... 183

Syllabus ... 183

sp^2 hybridization – ethene (C_2H_4) ... 185

sp hybridization – ethyne (C_2H_2) ... 187

Hybridization and molecular geometry ... 189

5.1 Measuring energy changes ... 191

Syllabus ... 191

Thermal energy (heat) transfer in chemical reactions ... 192

Open system ... 192

Closed system ... 192

Thermal energy (heat, q) ... 192

Enthalpy and enthalpy change ... 193

Standard enthalpy change of reaction ... 193

Standard conditions for enthalpy change ... 193

Endothermic and exothermic reactions ... 194

Enthalpy level diagrams ... 195

Standard enthalpy changes (Definitions) ... 196

Standard enthalpy change of reaction $\Delta H°_r$ 196
Standard enthalpy change of formation $\Delta H°_f$ 196
Standard enthalpy change of combustion $\Delta H°_c$ 196
Thermochemistry experiment – combustion of a fuel 198
Thermochemistry experiment – reaction in aqueous solutions 200
Thermochemistry experiment – neutralisation 202
5.2 Hess's law 205
    Hess's law 205
        Example 1 (ΔH_c from ΔH_f): 206
        Example 2 (ΔH_r from ΔH_c): 206
5.3 Bond enthalpy 209
    Syllabus 209
    Bond enthalpy (bond dissociation enthalpy), BE 209
        Bond enthalpy and bond length 210
        Bond enthalpy and multiple bonds 210
        Bond enthalpy and bond polarity 210
    Calculating Energy Changes 211
    Ozone (Trioxygen) 211
        Formation of Ozone 212
        (Natural) Destruction of Ozone 212
        (Anthropogenic) Destruction of Ozone 212
15.1 Energy cycles 213
    Syllabus 213
    Standard states 213
    Standard conditions 213
    Standard enthalpy changes 214
        Standard enthalpy change of formation $\Delta H°_f$ 214
        Lattice enthalpy (ΔH_{lat}) 214
        Enthalpy of atomisation (ΔH_{at}) 214
        Electron affinity (ΔH_{EA}) 215
    Born-Haber cycles 216
        Example 1: Calculating ΔH_f from a Born-Haber cycle for NaCl 216

Example 2: Calculating ΔH_f from a Born-Haber cycle for $MgCl_2$ 216

Example 3: Calculating ΔH_{lat} from a Born-Haber cycle for $CaCl_2$ 217

Example 4: Calculating ΔH_{lat} from a Born-Haber cycle for CaO 217

Lattice enthalpy trends – ionic radii 218

Lattice enthalpy trends – ionic charge 219

Solutions 220

Standard enthalpy change of solution (ΔH_{sol}) 220

Enthalpy change of hydration (ΔH_{hyd}) 220

15.2 Entropy and spontaneity 223

Syllabus 223

The Laws of Thermodynamics 223

Entropy change of a reaction (ΔS) 224

Gibbs free energy 226

Gibbs free energy – further explanation ($\Delta H > 0$ and $\Delta S > 0$) 226

Gibbs free energy – further explanation ($\Delta H > 0$ and $\Delta S < 0$) 226

Gibbs free energy – further explanation ($\Delta H < 0$ and $\Delta S > 0$) 227

Gibbs free energy – further explanation ($\Delta H < 0$ and $\Delta S < 0$) 227

Gibbs free energy change (ΔG) calculations 228

Gibbs free energy change of formation (ΔG_f) 229

Gibbs free energy change and extent of reversible reactions 230

6.1 Collision theory and rates of reaction 231

Syllabus 231

Rates of reaction 232

Concentration time graph 232

Concentration time graphs (typical examples) 234

Measuring rates of reaction – Gas evolved 236

Measuring rates – Loss of mass 238

Measuring rates of reaction – conductivity 238

Measuring rates – Colorimetry 239

Measuring rates of reaction – change in pH 239

Collision Theory 239

Criteria 1 – Energy of collision 240

- Criteria 2 – Geometry of collision ... 241
- Factors affecting reaction rates ... 241
  - Pressure ... 242
  - Concentration ... 243
  - Particle size (surface area) ... 244
  - Temperature ... 245
  - Catalyst ... 246
- 16.1 Rate expression and reaction mechanism ... 250
  - Syllabus ... 250
  - Rate equation ... 251
    - Units of rate constants ... 254
  - Concentration-time graphs for zero, first and second order reactions ... 255
  - Rate-concentration graphs for zero, first and second order reactions ... 256
- 16.2 Activation energy ... 261
  - Syllabus ... 261
  - Arrhenius equation ... 261
- 7.1 Equilibrium ... 263
  - Syllabus ... 263
  - Dynamic Equilibrium ... 264
    - Examples of physical systems in equilibrium ... 264
    - Examples of solutions in equilibrium ... 265
  - Reversible Reactions ... 266
  - The equilibrium law ... 267
  - Le Châtelier's principle ... 268
    - Equilibrium position ... 268
    - Le Châtelier's principle: Effect of concentration ... 269
    - Le Châtelier's principle: Effect of pressure ... 270
    - Le Châtelier's principle: Effect of temperature ... 271
    - Le Châtelier's principle: Effect of a catalyst ... 273
  - Reaction quotient ... 273
- 17.1 The equilibrium law ... 274
  - Syllabus ... 274

- Determining the equilibrium constant (the ICE method) 274
- Entropy and equilibrium 276
- 8.1 Theories of acids and bases 277
  - Syllabus 277
  - Preliminary concepts 278
    - Arrhenius Theory 278
    - Brønsted-Lowry Theory 279
- 8.2 Properties of acids and bases 281
  - Syllabus 281
  - General properties of acids 281
  - General properties of bases 282
  - Indicators 282
    - Acids and reactive metals 283
  - Acids and bases 284
    - Acids and bases (ionic equations) 284
  - Acids and carbonates 284
  - Acids and hydrogencarbonates 284
- 8.3 The pH scale 285
  - pH facts 285
    - The pH scale in context 285
  - Measuring pH 286
  - pH of strong acids 286
    - Neutrality of pure water 287
  - Ionic product constant 287
- 8.4 Strong and weak acids and bases 289
  - Syllabus 289
  - Strong acids 289
  - Weak acids 290
  - Strong bases 290
  - Weak bases 291
    - Differentiating strong and weak acids and bases 292
    - Differentiating between 'strength' and 'concentration' 293

- 8.5 Acid deposition 294
  - Syllabus 294
  - Definitions 295
  - Natural rainwater 295
  - Acid rain – effect of nitrogen oxides 295
  - Acid rain – effect of sulfur oxides 296
  - Measures to reduce acid rain pollutants 296
    - Pre-combustion methods 296
    - Post-combustion methods 296
- 18.1 Lewis acids and bases 298
  - Syllabus 298
  - Preliminary definitions 299
    - Examples of Lewis acids and bases 299
  - Quick review of pH 301
  - Acid dissociation constant, K_a 302
  - Base dissociation constant, K_b 302
  - Size of K_a, pK_a, K_b and pK_b 303
  - pH of weak acids 303
  - pH of weak bases 305
  - Conjugate acid-base pairs 306
    - Acid dissociation constant, K_a 306
    - Base dissociation constant, K_b 306
  - Ionic product of water, K_w (review) 306
- 9.1 Oxidation and reduction 308
  - Syllabus 308
  - Definition 1 (addition or removal of oxygen or hydrogen) 309
  - Definition 2 (loss or gain of electrons) 310
    - Common oxidising agents 311
    - Common reducing agents 311
  - Ionic half-equations 314
  - Oxidising agents which require an acid medium 314
    - Potassium manganate(VII) solution with dilute sulfuric acid. 314

- Dichromate(VI) ions in acid solution ... 314
- Overall redox equations from half equations ... 315
- Aqueous silver nitrate and copper metal ... 315
- Acidified dichromate (VI) and Fe^{2+} solution. ... 315
- Acidified potassium manganate (VII) and Sn2+ solution ... 315
- Half equations from overall redox equations ... 315
- Redox and oxidation numbers ... 316
- Reactivity – Redox couples ... 317
- Strength non-metals as oxidising agents ... 318
- Strength of metals as reducing agents ... 319
- The electrochemical series for metals ... 320
- Displacement reactions - metals ... 320
- Displacement reactions - halogens ... 320
- Redox titration ... 321
  - Redox titration with oxidising agents ... 321
  - Redox titration with reducing agents ... 322
- The Winkler method to determine Biochemical Oxygen Demand (BOD) ... 323
  - Introduction ... 323

9.2 Electrochemical cells ... 324
- Syllabus ... 324
- Preliminary definitions ... 325
- Half Cells ... 326
  - Zinc half cell ... 326
  - Copper half cell ... 326
- Voltaic Cells – Half Cells combined ... 327
  - Daniell cell ... 327
  - Other voltaic cells ... 328
- Electrolysis of molten sodium chloride (Downs cell) ... 329
- Electrolysis of molten lead bromide (carbon electrodes) ... 330
- Comparison between electrolytic and voltaic cells ... 331

19.1 Electrochemical cells ... 332
- Syllabus ... 332

Preliminary definitions... 333
Standard hydrogen electrode (SHE)... 333
Standard electrode potential ($E^\ominus$).. 334
    Measuring Standard Electrode Potential ($E^\ominus$) of metal half-cells...................... 334
    Measuring Standard Electrode Potential ($E^\ominus$) of gases 335
    Measuring Standard Electrode Potential ($E^\ominus$) of a mixture of ions 335
Spontaneity of Redox reactions 336
    Steps involved: 336
Electrolysis of aqueous solutions 339
    Electrolysis of aqueous solutions (ion selection) 341
Electrolysis of water using platinum electrodes 341
Electrolysis of concentrated brine (sodium chloride).. 342
Electrolysis of copper(II) sulfate solution using graphite electrodes 343
Electrolysis of copper(II) sulfate solution with copper electrodes 344
Electrolysis of sodium dicyanoargentate(I) solution – silver plating....................... 345
Quantitative electrolysis.. 346
    SI units 346
    Equations... 347
10.1 Fundamentals of organic chemistry 348
    Syllabus 348
    Introduction 349
        The carbon atom (electronic configuration $1s^2 2s^2 2p^2$) 349
    Bond geometry.. 350
    Homologous series.. 351
        Alkanes (general formula: C_nH_{2n+2}) 351
        Alkenes (general formula: C_nH_{2n})... 353
        Alcohols (general formula: $C_nH_{2n+1}OH$) 355
        Aldehydes (general formula: $C_nH_{2n}O$) 357
        Ketones (general formula: $C_nH_{2n}O$) 358
    Skeletal formulae 360
    Functional groups 361
    Nomenclature... 367

- Isomerism 368
  - Carbon Chain Isomerism 368
- Aromatic hydrocarbons 381
  - Aromatic hydrocarbons – evidence of benzene structure 382
- 10.2 Functional group chemistry 383
  - Syllabus 383
  - Alkanes – introduction 384
  - Alkanes – physical properties 384
  - Alkanes – Combustion 385
  - Alkanes – free-radical substitution (halogenation) 386
  - Alkenes – introduction 388
  - Alkenes – addition with hydrogen 388
  - Alkenes – addition with halogens 389
    - Alkenes – addition with halogens (mechanism) 389
  - Alkenes – addition with hydrogen halides 390
  - Alkenes – addition with water (steam) 390
  - Alkenes – addition polymerisation 391
  - Alcohols – introduction 392
  - Alcohols – combustion 392
  - Alcohols – oxidation of primary alcohols to aldehydes (mild conditions) 393
  - Alcohols – further oxidation of primary alcohols to carboxylic acids (strong conditions – reflux) 393
  - Alcohols – oxidation of secondary alcohols to ketones (strong conditions – reflux) 393
  - Alcohols – oxidation of tertiary alcohols 393
  - Alcohols – condensation with carboxylic acid to form esters 394
  - Introduction to mechanisms 396
    - Homolytic fission – formation of radicals 396
    - Electrophiles 396
    - Nucleophiles 397
  - Introduction to mechanisms – nucleophilic substitution 397
  - Introduction to mechanisms – electrophilic substitution 397
- 20.1 Types of organic reactions 398

Syllabus 398
Nucleophilic substitution 400
    The hydroxide ion, OH⁻ as a nucleophile 400
    Rules for drawing chemical mechanisms 400
    Nucleophilic substitution – S_N2 (a one-step reaction for primary halogenoalkanes) 401
    Nucleophilic substitution – S_N1 (a two-step reaction for tertiary halogenoalkanes) 402
    Nucleophilic substitution – factors determining the rate of reaction 403
    Electrophilic addition 405
    Electrophilic addition – alkene and halogen reactions 405
    Electrophilic addition – alkene and hydrogen halide reactions 406
        Markovnikov's Rule 406
    Electrophilic addition – alkene and interhalogen reactions 407
    Electrophilic substitution – nitration of benzene 408
    Electrophilic substitution – nitration of benzene 409
    Synthesis of phenylamine (aniline) from nitrobenzene 410
    Oxidation of alcohols and aldehydes (review) 411
    Nucleophilic addition – reduction of carboxylic acids, ketones and aldehydes 411
20.2 Synthetic routes 412
    Syllabus 412
    Synthesis of organic compounds 412
    Retro-synthesis 412
20.3 Stereoisomerism 414
    Syllabus 414
    Isomerism – a review 415
    Corformational Isomerism 415
    Configurational Isomerism – cis-trans and E-Z isomerism 417
        Planar molecules 417
        Cyclic molecules 418
    Configurational Isomerism – optical isomerism 419
11.1 Uncertainties and errors in measurement and results 425
    Syllabus 425
    Significant figures (SF) 426

Experimental errors – systematic errors 427

Experimental errors – random errors 428

Accuracy and Precision 428

Percentage error and accuracy 429

Uncertainty in measurement 429

    Comparison between accuracy and precision 430

Absolute, relative and percentage relative uncertainty 430

Propagation of uncertainties 431

Suggested improvements 434

Discrepancy from literature (accepted) value 434

11.2 Graphical techniques 435

    Syllabus 435

    Linear graphs 436

        Extrapolation of linear graphs 437

        Interpolation on curves 438

11.3 Spectroscopic identification of organic compounds 443

    Syllabus 443

    Index of Hydrogen Deficiency (IHD) 443

    Mass spectroscopy 444

    Infrared (IR) spectroscopy 446

        Infrared (IR) spectroscopy – the spring model and vibrational energy 447

        Infrared (IR) spectroscopy – spectrum analysis 447

        Infrared (IR) spectroscopy – progress of reaction 452

    ^{1}H NMR spectroscopy 453

21.1 Spectroscopic identification of organic compounds 456

    Syllabus 456

    High resolution ^{1}H NMR spectroscopy – spin-spin coupling 457

        High resolution ^{1}H NMR spectroscopy – the (n+1) rule and Pascal's triangle ... 459

        High resolution ^{1}H NMR spectroscopy – sample analysis 460

        High resolution ^{1}H NMR spectroscopy – TMS as the solvent 461

    X-ray crystallography 461

Introduction to Advanced Chemistry

Properties of matter

Physical properties
Properties which can be observed without changing the identity of the substance e.g. *melting, compressing, bending* and *magnetising*.

Chemical properties
Properties which can only be observed by changing the identity of the substance e.g. *burning, reaction with acid* and *rusting*.

Basic definitions
An **element**: a substance which consists of only **one type of atom** e.g. *iron, chlorine, carbon* and *neon*. It cannot be broken down into more substances.
A **compound**: a combination of two or more elements which are **chemically bonded** e.g. *salt (sodium chloride), lime (calcium hydroxide)* and *water*.
A **mixture**: a combination of elements and/or compounds which are not chemically bonded e.g. *sugary water* and *sparkling water*. These can be **physically separated**.

Fundamental forces
There are 4 fundamental forces in the universe:
Strong nuclear force and **weak nuclear force** are at work inside the nucleus of an atom. They keep the neutrons and protons together.
Gravitational force exists between all masses but the mass has to be very large to be significant.
Electromagnetic force exists between charged particles. In chemistry, these are called ions.
- Particles with **like charges repel** each other i.e. cations repel cations.
- Particles with **opposite charges attract** each other i.e. cations attract anions.

The **size of the forces** depends on:
- size of the **charges**.
- **distance** between the centre of the particles.

Energy changes
Atoms, molecules and ions are held together by electrostatic forces. When these forces are strong enough, they form **chemical bonds**.

Particles are brought closer together by these forces: chemical bonds are **formed**.
Energy is released as an **EXOTHERMIC** reaction.

Particles are move further apart from these forces: chemical bonds are **broken**.
Energy is absorbed as an **ENDOTHERMIC** reaction.

The chemical energy is usually converted into **heat (thermal)** energy. This is called the **enthalpy** of the chemical reaction.

Chemical calculations (introduction and review)

Units

Mass
In chemistry, the unit for mass is usually the **gram (g)**.
- 1 g = 1000 mg (1 mg = 10^{-3} g)
- 1 g = 1 000 000 g (1 μg = 10^{-6} g)
- 1 g = 1 000 000 000 ng (1 ng = 10^{-9} g)
- 1 kg = 1000 g
- 1 tonne = 1000 kg

Volume
In chemistry, the units for volume are usually **cm³** (small volumes) or **dm³** (large volumes).
- 1 dm³ (1 litre) = 1000 cm³
- 1 cm³ = 0.001 dm³ = 10^{-3} dm³

1.1 Introduction to the particulate nature of matter and chemical change

Syllabus

Nature of science – can you relate this topic to these concepts?
Making quantitative measurements with replicates to ensure reliability – definite and multiple proportions.

Understandings – how well can you explain these statements?
Atoms of different elements combine in fixed ratios to form compounds, which have different properties from their component elements.
Mixtures contain more than one element and/or compound that are not chemically bonded together and so retain their individual properties.
Mixtures are either homogeneous or heterogeneous.

Applications and skills – how well can you do all of the following?
Deduction of chemical equations when reactants and products are specified.
Application of the state symbols (s), (l), (g) and (aq) in equations.
Explanation of observable changes in physical properties and temperature during changes of state.

Matter

Traditionally called 'substance', matter is made up of **particles** (atoms, molecules or ions).
These particles are in **constant motion** at temperatures above 0 K (absolute zero).
All matter have **mass** and occupy a **volume** in space.

States of matter

Solid	Liquid	Gas
The particles are packed closely together. Interparticle forces are strong enough so that the particles cannot move freely but can only vibrate about a fixed position. A solid has a fixed shape and a fixed volume. Solids cannot be compressed.	The particles are very close together. Interparticle forces are weaker so that the particles have enough energy to move relative to each other in a fixed volume. The shape of a liquid is not definite but is determined by its container. Liquids cannot be compressed.	The particles have enough kinetic energy so that the effect of interparticle forces is negligible (or zero for an ideal gas). Gases have no definite shape or volume but spread out to fill all the available space. Gases can be compressed.

Kinetic Theory of Matter
- All matter consists of particles in motion. As the **temperature** increases, the **movement** of the particles increases.
- When the temperature increases enough for the particles to overcome the **interparticle forces**, a **change of state** occurs.

This is also true when the temperature decreases enough for the particles to exert interparticle forces on each other.

Solids can be transformed into liquids by **melting**, and liquids can be transformed into solids by **freezing**. Solids can also change directly into gases through the process of **sublimation**.

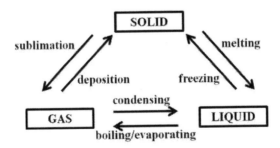

The kelvin scale for temperature

Absolute zero or zero kelvin (0 K)

This is the temperature at which the movement of particles completely stop.

The temperature in kelvin is **proportional to the average kinetic energy** of the particles.

To convert from degree celsius to kelvin, add 273.15.

T (K) = T (°C) + 273.15
T (°C) = T (K) − 273.15

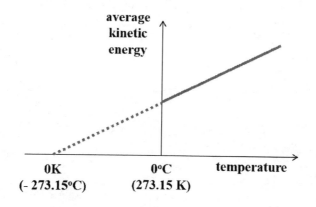

State symbols

Element	Atom or molecule	Formula
Argon	Atom	Ar
Iron	Atom	Fe
Hydrogen	Molecule	H_2
Sulfur	Molecule	S_8

Physical state	State symbol	Examples
Solid	(s)	Mg (s), $CaCO_3$(s), MnO_2 (s)
Liquid (pure liquid or molten)	(l)	H_2O (l), C_2H_5OH (l), C_8H_{18} (l), Fe (l)
Gas	(g)	H_2 (g), HCl (g), NH_3 (g), C_2H_6 (g)
Aqueous solution (dissolved in water)	(aq)	NaOH (aq), HCl (aq), KNO_3 (aq)

Pure substance - Elements
An element is made up of only one type of atom – it can be made up of single atoms or molecules (two or more atoms chemically bonded together) e.g.

Pure substance - Compounds
A compound is made up of two or more different types of atoms chemically bonded together – the bond can be ionic or covalent e.g.

Compound	Ionic or covalent	Formula
Magnesium chloride	Ionic	$MgCl_2$
Sulfuric acid	Covalent	H_2SO_4
Calcium carbonate	Ionic and covalent	$CaCO_3$

Mixtures

Pure substances (elements and compounds) can be combined to make mixtures. The substances can usually be separated by physical methods such as filtration, distillation or chromatography.

Homogeneous mixtures

These have uniform composition and properties throughout the mixture. The substances that make up the mixture are miscible with one another. Examples: salt water, metal alloys, vodka.

Heterogeneous mixtures

These have non-uniform composition and varying properties throughout the mixture. The substances that make up the mixture are immiscible with one another. Examples: a bar of chocolate with nuts, salad dressing, a bowl of cereal in milk.

Ions

Ions are charged species.

Positive ions (**cations**) are formed when a substance has lost electrons.

$$A \rightarrow A^{n+} + ne^- \quad \text{OR} \quad A - ne^- \rightarrow A^{n+}$$

Negative ions (**anions**) are formed when a substance gains electrons.

$$B + ne^- \rightarrow B^{n-}$$

Cations and their charges

	1+ charge	2+ charge	3+ charge
Group 1	Li^+, Na^+, K^+, Rb^+, Cs^+		
Group 2		Be^{2+}, Mg^{2+}, Ca^{2+}, Sr^{2+}, Ba^{2+}	
Group 3			Al^{3+}
d-block	Ag^+ Cu^+ (in copper (I) compounds)	Zn^{2+} Mn^{2+} (in manganese (II) compounds Fe^{2+} (in iron (II) compounds) Cu^{2+} (in copper (II) compounds)	Cr^{3+} (in chromium (III) compounds) Fe^{3+} (in iron (III) compounds)
Non-metal	NH_4^+ (ammonium)		

Anions and their charges

	1- charge	2- charge	3- charge
Group 7	F^- (fluoride) Cl^- (chloride) Br^- (bromide) I^- (iodide)		
Group 6		O^{2-} (oxide) S^{2-} (sulfide)	
Group 5			N^{3-} (nitride)
Polyatomic	OH^- (hydroxide) NO_3^- (nitrate) HCO_3^- (hydrogencarbonate) MnO_4^- (manganate (VII)) OCl^- (chlorate (I)) ClO_3^- (chlorate (V)) CN^- (cyanide) O_2^- (superoxide)	O_2^{2-} (peroxide) CO_3^{2-} (carbonate) SO_4^{2-} (sulfate) SO_3^{2-} (sulfite) CrO_4^{2-} (chromate (VI)) $Cr_2O_7^{2-}$ (dichromate (VI))	PO_4^{3-} (phosphate) PO_3^{3-} (phosphonate)

Formulae of ionic compounds

In ionic compounds the total number of positive charges equals the total number of negative charges.

Name	Cation	Anion	Ratio to balance charges	Formula
Sodium hydroxide	Na^+	OH^-	1:1	$NaOH$
Aluminium nitride	Al^{3+}	N^{3-}	1:1	AlN
Iron (II) sulfate	Fe^{2+}	SO_4^{2-}	1:1	$FeSO_4$
Aluminium oxide	Al^{3+}	O^{2-}	2:3	Al_2O_3
Calcium hydrogencarbonate	Ca^{2+}	HCO_3^-	1:2	$Ca(HCO_3)_2$
Sodium carbonate	Na^+	CO_3^{2-}	2:1	Na_2CO_3
Aluminium sulfate	Al^{3+}	SO_4^{2-}	2:3	$Al_2(SO_4)_3$
Ammonium sulfate	NH_4^+	SO_4^{2-}	2:1	$(NH_4)_2SO_4$

Common acids

Name of acid	Formula	Name of acid	Formula
Hydrochloric acid	HCl	Phosphoric(V) acid	H_3PO_4
Nitric(V) acid	HNO_3	Methanoic acid	$HCOOH$
Sulfuric(VI) acid	H_2SO_4	Ethanoic acid	CH_3COOH

Balancing chemical equations

Steps to balance chemical equations:
1. Select the element that appears in the fewest number of species in the equation. Balance it first.
2. If there is more than one of these, start with the one with the least number of atoms in the formula.
3. Balance the next element using steps 1-2.
4. Polyatomic ions e.g. NH_4^+, SO_4^{2-}, NO_3^-, etc. should be considered as entities (do not separate them into atoms).
5. Include the relevant state symbols.

Example 1
Calcium carbonate reacts with dilute nitric acid to form calcium nitrate, carbon dioxide and water.
$CaCO_3$ (s) + HNO_3 (aq) → $Ca(NO_3)_2$ (aq) + CO_2 (g) + H_2O (l)
$HNO_3 \times 2$:
$CaCO_3$ (s) + 2HNO_3 (aq) → $Ca(NO_3)_2$ (aq) + CO_2 (g) + H_2O (l)

Example 2
Copper (II) sulfate solution reacts with sodium hydroxide to form a precipitate of copper (II) hydroxide and a solution of sodium sulfate.
$CuSO_4$ (aq) + NaOH (aq) → $Cu(OH)_2$ (s) + Na_2SO_4 (aq)
NaOH×2:
$CuSO_4$ (aq) + 2NaOH (aq) → $Cu(OH)_2$ (s) + Na_2SO_4 (aq)

Example 3
Aluminium reacts with dilute sulfuric acid to form aluminium sulfate and hydrogen.
Al (s) + H_2SO_4 (aq) → $Al_2(SO_4)_3$ (aq) + H_2 (g)
Al×2:
2Al (s) + H_2SO_4 (aq) → $Al_2(SO_4)_3$ (aq) + H_2 (g)
$H_2SO_4 \times 3$:
2Al (s) + 3H_2SO_4 (aq) → $Al_2(SO_4)_3$ (aq) + H_2 (g)
$H_2 \times 3$:
2Al (s) + 3H_2SO_4 (aq) → $Al_2(SO_4)_3$ (aq) + 3H_2 (g)

Example 4
Propane, C_3H_8 burns in excess air to form carbon dioxide and water.
C_3H_8 (g) + O_2 (g) → CO_2 (g) + H_2O (l)
$CO_2 \times 3$:
C_3H_8 (g) + O_2 (g) → 3CO_2 (g) + H_2O (l)
$H_2O \times 4$:
C_3H_8 (g) + O_2 (g) → 3CO_2 (g) + 4H_2O (l)
$O_2 \times 5$:
C_3H_8 (g) + 5O_2 (g) → 3CO_2 (g) + 4H_2O (l)

Example 5
Ethane, C_2H_6 burns in excess air to form carbon dioxide and water.
C_2H_6 (g) + O_2 (g) → CO_2 (g) + H_2O (l)
$CO_2 \times 2$:
C_2H_6 (g) + O_2 (g) → $2CO_2$ (g) + H_2O (l)
$H_2O \times 3$:
C_2H_6 (g) + O_2 (g) → $2CO_2$ (g) + $3H_2O$ (l)
$O_2 \times 7/2$:
C_2H_6 (g) + 7/2O_2 (g) → $2CO_2$ (g) + $3H_2O$ (l)
The equation can also be multiplied by 2 to give:
$2C_2H_6$ (g) + $7O_2$ (g) → $4CO_2$ (g) + $6H_2O$ (l)

Ionic equations
1. Write the full balanced equation including the state symbols.
2. Separate the ions for dissolved (aqueous) ionic substances.
3. Keep the full formula intact for solids, liquids and gases.
4. Cross out the spectator ions (ones that appear on both sides of the equation).

Example 1
Copper (II) sulfate solution reacts with sodium hydroxide to form a precipitate of copper (II) hydroxide and a solution of sodium sulfate.
$CuSO_4$ (aq) + NaOH (aq) → $Cu(OH)_2$ (s) + Na_2SO_4 (aq)
NaOH×2:
$CuSO_4$ (aq) + 2NaOH (aq) → $Cu(OH)_2$ (s) + Na_2SO_4 (aq)
Cu^{2+} (aq) + ~~SO_4^{2-} (aq)~~ + $2Na^+$ ~~(aq)~~ + $2OH^-$ (aq) → $Cu(OH)_2$ (s) + ~~$2Na^+$ (aq)~~ + ~~SO_4^{2-} (aq)~~
Cu^{2+} (aq) + $2OH^-$ (aq) → $Cu(OH)_2$ (s)

Example 2
Magnesium reacts with dilute hydrochloric acid to form magnesium chloride solution and hydrogen gas.
Mg (s) + HCl (aq) → $MgCl_2$ (aq) + H_2 (g)
HCl×2:
Mg (s) + 2HCl (aq) → $MgCl_2$ (aq) + H_2 (g)
Mg (s) + $2H^+$ (aq) + ~~$2Cl^-$ (aq)~~ → Mg^{2+} (aq) + ~~$2Cl^-$ (aq)~~ + H_2 (g)
Mg (s) + $2H^+$ (aq) → Mg^{2+} (aq) + H_2 (g)

Example 3
Aluminium oxide reacts with sulfuric acid to form aluminium sulfate solution and water.
Al_2O_3 (s) + H_2SO_4 (aq) → $Al_2(SO_4)_3$ (aq) + H_2O (l)
$H_2SO_4 \times 3$:
Al_2O_3 (s) + $3H_2SO_4$ (aq) → $Al_2(SO_4)_3$ (aq) + H_2O (l)
$H_2O \times 3$:
Al_2O_3 (s) + $3H_2SO_4$ (aq) → $Al_2(SO_4)_3$ (aq) + $3H_2O$ (l)
Al_2O_3 (s) + $6H^+$ (aq) + ~~$3SO_4^{2-}$ (aq)~~ → $2Al^{3+}$ (aq) + ~~$3SO_4^{2-}$ (aq)~~ + H_2O (l)
Al_2O_3 (s) + $6H^+$ (aq) → $2Al^{3+}$ (aq) + H_2O (l)

Solubility of ionic compounds

Most ionic compounds are soluble in water according to these rules:
Soluble:
- All group 1 metal compounds.
- All ammonium compounds.
- All nitrate compounds.
- All chloride compounds except silver chloride and lead (II) chloride.
- All sulfate compounds except strontium sulfate, barium sulfate and lead (II) sulfate. Calcium sulfate is very slightly soluble.

Insoluble:
- All carbonate compounds except group 1 and ammonium carbonates.
- All hydroxide compounds except group 1, ammonium and barium hydroxides. Calcium and strontium hydroxides are slightly soluble.

Atom Economy

$$\% \text{ atom economy} = \frac{\text{molecular mass of atoms of useful products}}{\text{molecular mass of atoms in reactants}} \times 100$$

Calculate the atom economy for the following industrial processes:
Producing hydrogen from reacting coal with steam.
C (s) + $2H_2O$ (g) → CO_2 (g) + $2H_2$ (g)
Reducing iron oxide to iron in a blast furnace
Fe_2O_3 (s) + 3CO (g) → 2Fe (l) + $3CO_2$ (g)
Fermentation of sugar (glucose) to produce alcohol
$C_6H_{12}O_6$ (aq) → $2C_2H_5OH$ (aq) + $2CO_2$ (g)

Producing hydrogen from reacting natural gas with steam
$$CH_4\ (g) + H_2O\ (g) \rightarrow 3H_2\ (g) + CO\ (g)$$

1.2 The mole concept

Syllabus
Nature of science – can you relate this topic to these concepts?
Concepts – the concept of the mole developed from the related concept of "equivalent mass" in the early 19th century.
Understandings – how well can you explain these statements?
The mole is a fixed number of particles and refers to the amount, n, of substance.
Masses of atoms are compared on a scale relative to ^{12}C and are expressed as relative atomic mass (A_r) and relative formula/molecular mass (M_r).
Molar mass (M) has the units g mol^{-1}.
The empirical formula and molecular formula of a compound give the simplest ratio and the actual number of atoms present in a molecule respectively.
Applications and skills – how well can you do all of the following?
Calculation of the molar masses of atoms, ions, molecules and formula units.
Solution of problems involving the relationships between the number of particles, the amount of substance in moles and the mass in grams.
Interconversion of the percentage composition by mass and the empirical formula.
Determination of the molecular formula of a compound from its empirical formula and molar mass.
Obtaining and using experimental data for deriving empirical formulas from reactions involving mass changes.

The mole and the Avogadro constant
Imagine this:
1 g of water contains 33 400 000 000 000 000 000 000 water molecules. To avoid calculating with these large numbers, the **mole** concept was introduced.
One **mole** is the **amount of substance containing the Avogadro number of atoms, molecules or ions.**
Avogadro number (L or N_A) = 6.02×10^{23} mol^{-1}
So
One mole is the amount of substance containing 6.02×10^{23} mol^{-1} of atoms, molecules or ions e.g.
1 mol of water (H_2O) contains 6.02×10^{23} mol^{-1} water molecules.
1 mol of sodium (Na) contains 6.02×10^{23} mol^{-1} sodium atoms.
1 mol of sodium chloride (NaCl) contains 6.02×10^{23} mol^{-1} sodium chloride molecules or 6.02×10^{23} mol^{-1} pairs of Na^+ and Cl^- ions.

Mole ratios
Chemicals react according to a ratio by moles (not always 1:1).
$AgNO_3 + NaCl \rightarrow AgCl + NaNO_3$
1 mole of silver nitrate reacts with 1 mole of sodium chloride to give 1 mole of silver chloride and 1 mole of sodium nitrate.

$H_3PO_4 + 3KOH \rightarrow K_3PO_4 + 3H_2O$
1 mole of phosphoric acid reacts with 3 moles of potassium hydroxide to give 1 mole of potassium phosphate and 3 moles of water.

$2Al + 3H_2SO_4 \rightarrow Al_2(SO_4)_3 + 3H_2$
2 moles of aluminium reacts with 3 moles of sulfuric acid to give 1 mole of aluminium sulphate and 3 moles of hydrogen.

$H_2SO_4 + 2NaOH \rightarrow Na_2SO_4 + 2H_2O$
1 mole of sulfuric acid reacts with 2 moles of sodium hydroxide to give 1 mole of sodium sulfate and 2 moles of water.
0.5 moles of sulfuric acid reacts with 1 moles of sodium hydroxide to give 0.5 mole of sodium sulfate and 1 moles of water.

Measurements and units

Systeme International (SI) units

Property	Symbol	Unit	Symbol
mass	m	kilogram	kg
time	t	second	s
temperature	T	kelvin	K
amount	n	mole	mol
electric current	I	ampere	A
luminosity	I_v	candela	cd
length	l	metre	m

Derived units

Property	Symbol	Unit	Symbol
volume	V	cubic metre	m^3
pressure	P	pascal	Pa or Nm^{-2}

Chemistry in the laboratory usually deals with **smaller quantities**, so these units are **more convenient**.

Property	Unit	Symbol
mass	gram	g
time	minute	min
temperature	degree Celcius	°C
volume	cubic centimetre or cubic decimetre	cm^3 or dm^3
pressure	atmosphere	atm

Prefixes for units

Prefix	pico	nano	micro	milli
Symbol	p	n	μ	m
Scale	10^{-12}	10^{-9}	10^{-6}	10^{-3}

Prefix	centi	desi
Symbol	c	d
Scale	10^{-2}	10^{-1}

Prefix	kilo	mega	giga	tera
Symbol	k	M	G	T
Scale	10^3	10^6	10^9	10^{12}

Important conversions

1 dm = 10 cm

1 dm³ = 1000 cm³

Important constants

Avogadro's number (N_A)
The number of atoms or molecules in 1 mole of a substance is a constant.
This is the **Avogadro's number, $N_A = 6.02 \times 10^{23}$ mol^{-1}**
Summary of units used by chemists:
- **Mass** measured in **grams (g)** or **kilograms (kg)**
- **Volume** measured in **cubic centimetres (cm³)** or **cubic decimetres (dm³)**
- **Amount** measured in **moles (mol)**

Molar volume of an ideal gas at 273 K and 100 kPa
  $= 2.27 \times 10^{-2}$ m³ mol^{-1} = 22.7 dm³ mol^{-1}

Ratio of elements
Chemical formulae tell us how many atoms of each element there are per formula. Atoms are too small to weigh individually.
ANY mass of water will have hydrogen and oxygen in the same ratio.
e.g.
18 kg of water contains 2 kg of hydrogen and 16 kg of oxygen.
180 tonnes of water contains 20 tonnes hydrogen and 160 tonnes of oxygen.

The mole
The number of atoms will be very large in 180 tonnes of water!
It is useful to have a unit of **AMOUNT** representing a specific number of particles e.g. atoms, molecules, ions and electrons. Chemists use the **MOLE** as a unit of amount.
A **mole** of substance is the **amount of substance** that has the **same number of particles** as there are **atoms in exactly 12 g of carbon-12**.
Compare this with other definitions:
- One mole is the amount of substance containing the Avogadro number of atoms, molecules, ions or electrons.
- One mole is the amount of substance containing 6.02×10^{23} mol^{-1} of atoms, molecules, ions or electrons.

They all mean the same thing.

Formulas

Molar mass
Molar mass is the mass of one mole of a substance (atomic element, molecular element or compound) in g mol^{-1}. The unit is not usually quoted e.g.
molar mass of sulphur, M_r (S) = 32.06 g mol^{-1}
molar mass of chlorine, M_r (Cl$_2$) = 70.90 g mol^{-1}
The mole is useful when measuring out reactants or calculating product mass in a reaction.

Relative atomic mass (A_r)

The **relative atomic mass** of an atom is the weighted average of the atomic masses of its isotopes and their relative abundances.

Chemists can count atoms and molecules by weighing them because **different elements have different masses.**

John Dalton (19th C) was the first chemist to use the name 'atom' for the smallest particle of an element.

Relative atomic masses of some elements:

Element	Symbol	A_r
Carbon-12	^{12}C	12.00
Carbon	C	12.01
Chlorine	Cl	35.45
Copper	Cu	63.55
Hydrogen	H	1.01
Iron	Fe	55.85
Magnesium	Mg	24.31
Sulphur	S	32.07

The mass of one atom of **carbon-12 isotope** is defined as being **exactly 12.00**.

Examples of weighted average:

Isotope	Relative abundance (%)	Atomic mass
^{32}S	96.1	32.0
^{33}S	0.8	33.0
^{34}S	3.1	34.0
A_r (sulfur)		**32.07**

Isotope	Relative abundance (%)	Atomic mass
^{24}Mg	79.4	24.0
^{25}Mg	10.0	25.0
^{26}Mg	10.6	26.0
A_r (magnesium)		**24.31**

Relative molecular mass (M_r)

Masses of different molecules can also be compared, to give the relative molecular mass (r.m.m.) or M_r which is the **mass of a molecule of the compound divided by 1/12 the mass of a carbon-12 atom.**

To calculate the relative molecular mass for any compound, it is a simple matter of adding up the individual relative atomic masses of the atoms in the molecule.

Examples of relative molecular mass:

$M_r (CH_4) = (1 \times 12.01) + (4 \times 1.01) = 16.05$

$M_r (HNO_3) = (1 \times 1.01) + (1 \times 14.01) + (3 \times 16.00) = 63.02$

Relative formula mass

When compounds contain **ions** rather than molecules, the term relative formula mass is used. The calculation method is the same as for M_r using the number of ions present in the formula of a compound.

Examples of relative formula mass:

$M_r (Na_2O) = (2 \times 22.99) + (1 \times 16.00) = 61.98$

$M_r (CuCl_2) = (1 \times 63.55) + (2 \times 35.45) = 134.45$

$M_r (SO_4^{2-}) = (1 \times 32.06) + (4 \times 16.00) = 96.06$

Calculating number of moles from mass
To calculate the amount of a substance in moles, divide the mass used by the molar mass.
Amount of substance (number of moles) = Mass / Molar mass
n = m / M
Example:
What is the amount in moles of 584.4 g of Na Cl?
Amount of NaCl = 584.4 / 58.44 = 10 mol

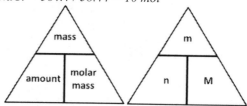

Calculating the mass of substance
To calculate the mass of a substance from an amount in moles, just rearrange the previous formula for calculating amount.
Mass = Amount × Molar mass
Examples:
Calculate the mass of the following:
a) 0.1 mol of carbon dioxide
   mass of CO_2 = 0.1 × (12.01 + 2 × 1.01) = 1.40 g
b) 10 mol of calcium carbonate, $CaCO_3$
   mass of $CaCO_3$ = 10 × (40.08 + 12.01 + 3 × 16.00) = 1000.90 g

Percentage composition

$$\% \text{ composition} = \frac{\Sigma A_r \text{ of an element} \times 100}{M_r \text{ of compound}}$$

Calculate the % composition of the stated element in each of the following:

C_2H_4 (carbon)

$$\% \text{ carbon} = \frac{2 \times 12.01}{2 \times 12.01 + 4 \times 1.01} \times 100 = 85.60\,\%$$

Na_2O (sodium)

$$\% \text{ sodium} = \frac{2 \times 22.99}{2 \times 22.99 + 16.00} \times 100 = 74.19\,\%$$

$C_6H_{12}O_6$ (carbon)

$$\% \text{ carbon} = \frac{6 \times 12.01}{6 \times 12.01 + 12 \times 1.01 + 6 \times 16.00} \times 100 = 39.99\,\%$$

K_2SO_4 (oxygen)

$$\% \text{ oxygen} = \frac{4 \times 16.00}{2 \times 39.10 + 32.06 + 4 \times 16.00} \times 100 = 36.73\,\%$$

Empirical and molecular formulae

The **empirical formula** of a substance shows the **simplest whole number ratio** of the atoms of each element present.

The **molecular formula** of a substance shows the **number of atoms of each element** present in one molecule of the substance.

In many cases the empirical formula is the same as the molecular formula.

Examples:

Compound	Molecular	Empirical
Water	H_2O	H_2O
Methane	CH_4	CH_4
Butane	C_4H_{10}	C_2H_5
Hydrogen peroxide	H_2O_2	HO
Octane	C_8H_{18}	C_4H_9
Benzene	C_6H_6	CH
Glucose	$C_6H_{12}O_6$	CH_2O

Calculating empirical and molecular formulae

Empirical formulae can be calculated from the percentage composition by mass of a compound. Experimental methods are used to determine the mass of each element in a compound.

Procedure:
1. Divide the % mass or mass by the A_r of the element.
2. Divide the values from step 1 with the smallest number.
3. If the numbers in step 2 are not whole numbers, multiply them by 2, 3, …

Example 1

Calculate the empirical formula for magnesium oxide when magnesium is weighed before and after burning in excess oxygen. Typical results might be:

0.560 g of magnesium produces **0.960 g** of magnesium oxide
Mass of oxygen = 0.960 - 0.560 = 0.400 g

	Mg	O
Mass	0.560	0.400
÷ A_r	0.560/24.13 = 0.023	0.400/16.00 = 0.025
÷ 0.023	= 1	= 1.09 ≈ 1
Ratio	1	1

Empirical formula is **MgO**

Example 2

An oxide of copper has the following composition by mass:
Copper 0.635 g ; Oxygen 0.080 g
Calculate the empirical formula of the oxide.

	Cu	O
Mass	0.635	0.080
÷ A_r	0.635/63.55 = 0.010	0.080/16.00 = 0.005
÷ 0.005	= 2	= 1
Ratio	2	1

Empirical formula is **Cu_2O**

Example 3
A compound contains 38.7% calcium, 20.0% phosphorus and 41.3 % oxygen by mass. Its molar mass is 310.18 g mol^{-1}.
Calculate the empirical and molecular formula of the compound.

	Ca	P	O
% mass	38.7	20.0	41.3
$\div A_r$	38.7/40.08 = 0.97	20.0/30.97 = 0.65	41.3/16.00 = 2.58
$\div 0.65$	= 1.5	= 1	= 4
x 2	= 3	= 2	= 8
Ratio	3	2	8

Empirical formula is **Ca$_3$P$_2$O$_8$**.
Empirical mass = (3×40.08) + (2×30.97) + (8×16.00) = 310.18
Molar mass = Empirical mass
Molecular formula is **Ca$_3$P$_2$O$_8$**

Example 4
A compound contains 34.3% sodium, 17.9% carbon and 47.8% oxygen by mass. Its molar mass is 134.00 g mol^{-1}.
Calculate the empirical and molecular formula of the compound.

	Na	C	O
% mass	34.3	17.9	47.8
$\div A_r$	34.3/22.99 = 1.49	17.9/12.01 = 1.49	47.8/16.00 = 2.99
$\div 1.49$	= 1	= 1	= 2.01 ≈ 2
Ratio	1	1	2

Empirical formula is **NaCO$_2$**
Empirical mass = (1×22.99) + (1×12.01) + (2×16.00) = 67.00
Molar mass = Empirical mass×2
Molecular formula is **Na$_2$C$_2$O$_4$**

Example 5
A technique known as **combustion analysis** is used to find the composition by mass of organic compounds. It involves the **complete combustion in excess oxygen** giving only **carbon dioxide** and **water** as products.

0.500 g of organic compound X (containing C, H, O only) produces 0.733 g CO_2 and 0.300 g H_2O.
Its mass spectrum showed a molar mass of 60.06 g mol^{-1}.
Determine the molecular formula of the compound.

Mass of carbon = 12.01 / (12.01 + 2 × 16.00) × 0.733 = 0.200 g
Mass of hydrogen = (2 × 1.01) / (2 × 1.01 + 16.00) × 0.300 = 0.034 g
Mass of oxygen = 0.500 - 0.200 - 0.034 = 0.266 g

	C	H	O
mass	0.200	0.034	0.266
÷ A_r	0.200/12.01 = 0.017	0.034/1.0 = 0.034	0.266/16.00 = 0.017
÷ 0.017	= 1	= 2	= 1
Ratio	1	2	1

Empirical formula is **CH_2O**
Empirical mass = (1×12.01) + (2×1.01) + (1×16.00) = 30.03
Molar mass = Empirical mass x 2
Molecular formula is **$C_2H_4O_2$**

1.3 Reacting masses and volumes

Syllabus

Nature of science – can you relate this topic to these concepts?
Making careful observations and obtaining evidence for scientific theories – Avogadro's initial hypothesis.

Understandings – how well can you explain these statements?
Reactants can be either limiting or excess.
The experimental yield can be different from the theoretical yield.
Avogadro's law enables the mole ratio of reacting gases to be determined from volumes of the gases.
The molar volume of an ideal gas is a constant at specified temperature and pressure.
The molar concentration of a solution is determined by the amount of solute and the volume of solution.
A standard solution is one of known concentration.

Applications and skills – how well can you do all of the following?
Solution of problems relating to reacting quantities, limiting and excess reactants, theoretical, experimental and percentage yields.
Calculation of reacting volumes of gases using Avogadro's law.
Solution of problems and analysis of graphs involving the relationship between temperature, pressure and volume for a fixed mass of an ideal gas.
Solution of problems relating to the ideal gas equation.
Explanation of the deviation of real gases from ideal behaviour at low temperature and high pressure.
Obtaining and using experimental values to calculate the molar mass of a gas from the ideal gas equation.
Solution of problems involving molar concentration, amount of solute and volume of solution.
Use of the experimental method of titration to calculate the concentration of a solution by reference to a standard solution.

Calculating reacting masses

Stoichiometry is the quantitative relationship in moles between the **reactants** and **products** in a chemical reaction.

Limiting reactant: the substance that determines the theoretical yield (mass of product calculated from the chemical equation, assuming that all the reactant is converted into the product) of product in a reaction.

In any chemical reaction:
- The reagent that is completely used up is the **limiting reagent (L)**.
- The reagent that has some left over is said to have been **in excess (E)**.

$E + L \rightarrow P$

At the end of this reaction:
- L is completely used up, there is some E left over.

Example 1
How many moles of iodine are obtained from 1/6 mol of potassium iodate(V) in this reaction?

Equation is:
$KIO_3 (aq) + 5KI (aq) + 6H^+ (aq) \rightarrow 3I_2 (aq) + 6K^+ (aq) + 3H_2O (l)$

From the balanced equation, 1 mol of potassium iodate(V) produces 3 mol of iodine.

So, 1/6 mol of potassium iodate(V) produces $1/6 \times 3 = \frac{1}{2}$ mol of iodine.

Example 2
Calculate the mass of iron produced when 320.00 g of iron(III) oxide is reduced by carbon monoxide in a blast furnace.

$Fe_2O_3 (s) + 3CO (g) \rightarrow 2Fe (s) + 3CO_2 (g)$

Amount of Fe_2O_3 = mass/molar mass = $320.00 / (2 \times 55.85 + 3 \times 16.00)$ = 2.00 mol

From the balanced equation, 1 mol of Fe_2O_3 produces 2 mol of Fe

Amount of Fe = 4.00 mol

Mass of Fe = amount×molar mass = $4.00 \times 55.85 = 223.40$ g

Example 3
Sodium carbonate weighing 5.30 g is reacted with excess hydrochloric acid. Calculate the mass of sodium chloride produced.

$Na_2CO_3 (s) + 2HCl (aq) \rightarrow 2NaCl (aq) + CO_2 (g) + H_2O (l)$

Amount of Na_2CO_3 = $5.30 / (2 \times 22.99 + 12.00 + 3 \times 16.00) = 0.05$ mol

From the balanced equation, 1 mol of Na_2CO_3 produces 2 mol of NaCl

Amount of NaCl = $2 \times 0.05 = 0.10$ mol

Mass of NaCl = $0.10 \times (22.99 + 35.45) = 5.84$ g

Example 4
When potassium iodate (V) is allowed to react with acidified potassium iodide, iodine is formed.
KIO_3 (aq) + $5KI$ (aq) + $6H^+$ (aq) → $3I_2$ (aq) + $6K^+$ (aq) + $3H_2O$ (l)
What mass of KIO_3 is required to give 10.00 g of iodine?
From the balanced equation, 1 mol of KIO_3 produces 3 mol of I_2.
Amount of I_2 = 10.00 / (2×126.90) = 0.039 mol
Amount of KIO_3 required = 0.039 / 3 = 0.013 mol
Mass of KIO_3 = 0.013×(39.10 + 126.90 + 3×16.00) = 2.78 g

Example 5
Aspirin, $C_9H_8O_4$ is made by the reaction:
salicylic acid + ethanoic anhydride → aspirin + ethanoic acid
$C_7H_6O_3$ + $C_4H_6O_3$ → $C_9H_8O_4$ + $C_2H_4O_2$
How many grams of salicylic acid, $C_7H_6O_3$, are needed to make one aspirin tablet, which contains 0.33 g of $C_9H_8O_4$?
From the balanced equation, 1 mol of $C_7H_6O_3$ produces 1 mol of $C_9H_8O_4$.
Amount of $C_9H_8O_4$ = 0.33 / (9×12.01 + 8×1.01 + 4×16.00) = 0.0018 mol
Amount of $C_7H_6O_3$ required = 0.0018 mol
Mass of $C_7H_6O_3$ = 0.0018 × (7×12.01 + 6×1.01 + 3×16.00) = 0.25 g

Example 6
TNT (trinitrotoluene) is an explosive. The compound is made by the reaction:-
toluene + nitric acid → TNT + water
C_7H_8 (l) + $3HNO_3$ (l) → $C_7H_5N_3O_6$ (s) + $3H_2O$ (l)
Calculate the masses of toluene and nitric acid that must be used to make 10 tonnes of TNT. (1 tonne = 1000 kg)
Amount of TNT = 1 000 000 / (7×12.01 + 5×1.01 + 3×14.01 + 6×16.00) = 4402.38 mol
Amount of toluene required = 4 402.38 mol
Mass of toluene required = 4 402.38 ×(7×12.01 + 8×1.01) = 405 679.32 g = 405.68 kg = 0.406 tonnes
Amount of nitric acid required = 3×4402.38 = 13 207.14 mol
Mass of nitric acid required = 13 207.14×(1.01 + 14.01 + 3×16.00) = 832 313.96 g = 832.31 kg = 0.832 tonnes

Using masses of reactants to work out the balanced equation for a reaction.

We can use the mass of each substance taking part in a reaction to calculate the number of moles of each. The simplest ratio between these mole quantities will describe the equation.

Example 1
In a suitable organic solvent, tin metal reacts with iodine to give an orange solid and no other product. In such a preparation, 5.90 g of tin reacted completely with 25.40 g of iodine. What is the formula of the compound produced?
Amount of tin used = 5.90 / 118.69 = 0.050 mol
Amount of iodine (atoms) = 25.40 / 126.90 = 0.20 mol
Ratio of tin:iodine = 1:4
The compound has the empirical formula SnI_4. The molecular formula cannot be determined without the molar mass.

Example 2
Iron burns in chlorine to form iron chloride. An experiment showed that 5.60 g of iron combined with 10.65 g of chlorine. Deduce the equation for the reaction.
Amount of iron used = 5.60 / 55.85 = 0.10 mol
Amount of chlorine (atoms) = 10.65 / 35.45 = 0.30 mol
Ratio of iron:chlorine = 1:3
The compound has the empirical formula $FeCl_3$. Assuming that this is also the molecular formula:
Fe (s) + Cl_2 (g) → $FeCl_3$ (s)
Cl_2×3:
Fe (s) + 3Cl_2 (g) → $FeCl_3$ (s)
$FeCl_3$×2:
Fe (s) + 3Cl_2 (g) → 2$FeCl_3$ (s)
Fe×2:
2Fe (s) + 3Cl_2 (g) → 2$FeCl_3$ (s)

Example 3
A mass of 0.65 g of zinc powder was added to a beaker containing silver nitrate solution. When all the zinc had reacted, 2.16 g of silver were obtained.
Amount of zinc used = 0.65 / 65.38 = 0.01 mol
Amount of silver formed = 2.16 / 107.87 = 0.02 mol
Amount of silver produced by 1 mol of zinc = 0.02×(1 / 0.01) = 2 mol
The balanced equation is
Zn (s) + 2$AgNO_3$ (aq) → 2Ag (s) + $Zn(NO_3)_2$ (aq)

Percentage yield

The effectiveness of a synthetic process in chemistry is often determined by the percentage yield of the desired product. Reasons for not obtaining 100% yield include:
- Competing reactions.
- Handling or transfer losses.

$$\% \text{ yield} = \frac{\text{actual yield of desired product}}{\text{theoretical yield of desired product}} \times 100$$

Example
When 1000 g of sulfur dioxide is reacted with excess oxygen, 1225 g of sulfur trioxide is produced. What is the % yield?
$2SO_2 + O_2 \rightarrow 2SO_3$
Amount of SO_2 used = 1000 / (32.06 + 2×16.00) = 15.61 mol
Theoretical yield of SO_3 = 15.61 x (32.06 + 3 x 16.00) = 1258 g

$$\% \, yield = \frac{1225}{1258} \times 100 = 97.4\%$$

Calculations involving Gas Volumes

Gas Volumes

Avogadro's Law: Equal volumes of gases contain the same number of molecules under the same conditions of temperature and pressure.
- At **standard temperature and pressure (STP)** of **0°C** and **100 kPa**:
- 1 mole of any gas occupies **22 700 cm^3** or **22.7 dm^3**.

This is usually approximated for **'room temperature and pressure' (RTP)** to **24 dm^3**. RTP is **25°C (298 K)** and **100 kPa**.
Hence, 24 dm^3 of CO_2 and 24 dm^3 of N_2 both contain 1 mole of molecules at RTP.
For a gas,
number of moles = volume / molar volume
$n = V / V_{mol}$

Example 1
Sodium carbonate weighing 5.30 g is reacted with excess hydrochloric acid. Calculate the mass of sodium chloride produced and volume of carbon dioxide produced at room temperature and pressure.
Na_2CO_3 (s) + 2HCl (aq) → 2NaCl (aq) + CO_2 (g) + H_2O (l)
Amount of Na_2CO_3 = 5.30 / (2×22.99 + 12.01 + 3×16.00) = 0.050 mol
1 mol of Na_2CO_3 produces 2 mol of NaCl
Amount of NaCl = 2×0.050 = 0.10 mol
Mass of NaCl = 0.10×(22.99 + 35.45) = 5.84 g
1 mol of Na_2CO_3 produces 1 mol of CO_2
Amount of CO_2 = 0.050 mol
Volume of CO_2 = 0.050 x 24 = 1.2 dm^3

Example 2
50 cm^3 of propane was burnt in excess oxygen. What volume of oxygen reacts and what is the volume of carbon dioxide produced?
C_3H_8 (g) + $5O_2$ (g) → $3CO_2$ (g) + 4 H_2O (l)
Molar ratio of gases propane:oxygen:carbon dioxide = 1:5:3
Volume of propane = 50 cm^3
Volume of oxygen = 5 x 50 = 250 cm^3
Volume of carbon dioxide = 3 x 50 = 150 cm^3
N.B.
It is not necessary to use the molar volume of the gases here. The units for volume have been kept as cm^3.

Example 3
27.00 g of propane was burnt in excess oxygen. What volume of oxygen reacts and what is the volume of carbon dioxide produced?
$C_3H_8 (g) + 5O_2 (g) \rightarrow 3CO_2 (g) + 4 H_2O (l)$
Amount of propane = 27.00 / (3×12.01 + 8×1.01) = 0.61 mol
Amount of oxygen = 5 x 0.61 = 3.05 mol
Volume of oxygen = 3.05 x 24 = 73.2 dm^3
Amount of carbon dioxide = 3 x 0.61 = 1.83 mol
Volume of carbon dioxide = 1.83 x 24 = 43.92 cm^3

Example 4
Measurements showed that 20 cm^3 of hydrogen react with exactly 10 cm^3 of oxygen to form water at room temperature and pressure. What is the balanced equation for this reaction?
The ratio of reacting volumes is 20:10 or 2:1
Hence the reacting mole ratio of hydrogen:oxygen is also 2:1
So the balanced equation is:
$2H_2 (g) + O_2(g) \rightarrow 2H_2O(l)$

The gas laws

Boyle's law
Pressure of a gas is inversely proportional to its volume.
$P \propto 1/V$
$PV = k_1$
  where k_1 = constant
if amount of gas and temperature is constant

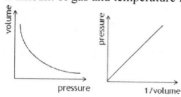

Charles' law
Volume of a gas is proportional to its temperature.
$V \propto T$
$V = k_2 T$
  where k_2 = constant
if amount of gas and pressure is constant

Gay-Lussac (Pressure) law
Pressure of a gas is proportional to its temperature.
$P \propto T$
$P = k_3 T$
  where k_3 = constant
if amount of gas and volume is constant

Combined gas law

$$\frac{PV}{T} = \text{constant}$$

$$\underbrace{\frac{P_1 V_1}{T_1}}_{\text{initial conditions}} = \underbrace{\frac{P_2 V_2}{T_2}}_{\text{final conditions}}$$

Ideal gas equation

Ideal gases are theoretical gases which are assumed to randomly moving and non-interacting. By assuming that the gas is ideal, the amount of gas, n can be included in the combined gas law to give the ideal gas equation.

$$\frac{PV}{RT} = n \quad \text{or} \quad PV = nRT$$

R = gas constant = 8.31 J K^{-1} mol^{-1}
n = amount of gas in mol
p = pressure in Pa (1 atm = 101 kPa = 1.01×10^5 Pa)
V = volume in m^3 (1 m^3 = 10^3 dm^3 = 10^6 cm^3)
T = temperature in K

For practical reasons, most gases can be assumed to be ideal gases if:
o the temperature is not too low.
o the pressure is not too high.

Example 1:
What is the volume occupied by 6.60 g of carbon dioxide gas at STP?
Amount of CO_2 = 6.60 / (12.01 + 2×16.00) = 0.15 mol
V = nRT / p = (0.15 × 8.31 × 273) / 1.01×10^5 = 3.37×10^{-3} m^3 = 3.37 dm^3

Example 2:
What is the amount of argon gas which occupies 42.15 dm^3 of space at STP? What is the mass of this gas?
n = pV / RT = (1.01×10^5 × 42.15×10^{-3}) / (8.31 × 273) = 1.88 mol
mass of Ar gas = 1.88 × 39.95 = 75.11 g

Concentrations of Solutions

$$\text{density of liquid} = \frac{\text{mass of liquid}}{\text{volume of liquid}} \qquad \rho = \frac{m}{V}$$

The density of pure water is **1 g cm^{-3}** or **1 kg dm^{-3}**.
In chemistry, aqueous solutions can be assumed to have the same density as water.
When dissolving mole quantities of a compound in a solvent, the resulting solution has its concentration expressed in mol dm^{-3}.

$$\text{concentration} = \frac{\text{amount of solute (mol)}}{\text{volume of solution (dm}^3\text{)}}$$

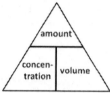

Concentration can also be quoted in **g dm^{-3}**.
Concentration (in g dm^{-3}) = concentration (in mol dm^{-3}) × M_r

Examples
5.84g of NaCl dissolved in water made up to make 1 dm^3 of solution has a concentration of 0.1 mol dm^{-3}
N.B.:
The term molar is often used instead of mol dm^{-3} e.g. 2 mol dm^{-3} of NaOH (aq) is 2M NaOH (aq) (or '2 molar'). This term should not be used.

Calculate the amount in moles of nitric acid in 25 cm^3 of 0.10 mol dm^{-3} aqueous solution.
Volume of solution (converted to dm^3) = 25 / 1000 dm^3
Amount of HNO$_3$ = 0.10×(25 / 1000) = 0.0025 mol

Calculate the concentration in mol dm^{-3} of aqueous solution containing 2.00×10^{-4} mol of sulfuric acid in 10.00 cm^3.
Concentration = 2.00×10^{-4} / (10.00 / 1000) = 2.00×10^{-8} mol dm^{-3}

Calculate the concentration in mol dm^{-3} of a solution comprising 0.125 mol of nitric acid with water added, up to a volume of 50.00 cm^3.
Concentration = 0.125 / (50.00 / 1000) = 2.5 mol dm^{-3}

What is the concentration in g dm^{-3} of 0.50 mol dm^{-3} aqueous ethanoic acid (CH$_3$COOH)?
Concentration = 0.50×(2×12.01 + 4×1.01 + 2×16.00) = 30.03 g dm^{-3}

What is the concentration in mol dm^{-3} of an aqueous solution containing 4.00 g dm^{-3} of sodium hydroxide?
Concentration = 4.00 / (22.99 + 16.00 + 1.01) = 0.10 mol dm^{-3}

Very small concentrations
For very small concentrations, the following units may be used:
- millimoles per dm^3 (mmol dm^{-3})
- micromoles per dm^3 (μmol dm^{-3})
- nanomoles per dm^3 (nmol dm^{-3})
- parts per million (ppm)
- parts per billion (ppb)

Example 1
The concentration of aluminium in drinking water is 6.3 nmol dm^{-3}. Calculate the mass of aluminium in 1 dm^3 of water.
Amount of aluminium in 1 dm^3 = 6.3 nmol
Mass of aluminium = 6.3×10^{-9}×26.92 = 1.70×10^{-7} g = <u>170 ng</u>

Example 2
A bottle of mineral water contains 2.0 ppm Na$^+$ ions. Calculate the concentration of Na$^+$ ions in nmol dm^{-3} and the number of Na$^+$ ions in 1.0 dm^3 of mineral water.
1 000 000 g of water contains 2.0 g of Na$^+$ ions.
1.0 dm^3 of water has a mass of 1000g.
1.0 dm^3 of water contains 2.0 ÷ (1 000 000/1000) = 0.0020 g Na$^+$ ions
Amount of Na$^+$ ions = 0.0020 / 22.99 = 8.70×10^{-5} mol = 8.70×10^4 nmol
Concentration of Na$^+$ ions = 8.70×10^4 nmol dm^{-3}
Number of Na$^+$ ions = 8.70×10^{-5}×6.02×10^{23} = 5.24×10^{19}

Titration calculations

Often a key stage in titration work is to calculate the amount (in moles) of reagent from a given concentration and volume.

There are 5 aspects to titration calculations:-
- A balanced equation giving mole ratios between reactants
- Volume of solution for first reagent
- Concentration of solution for first reagent
- Volume of solution for second reagent
- Concentration of solution for second reagent

Knowing 4 of these aspects allows you to work out the fifth.

Example 1
What amount of NaOH is present in 24 cm^3 of an aqueous solution with a concentration of 0.01 mol dm^{-3}?
Answer
Convert the volume to dm^3
volume of solution = 24 / 1000 dm^3
amount of NaOH = 0.01× 24/1000 = 2.4×10^{-4} mol

Example 2

A 25.00 cm^3 portion of sodium hydroxide is neutralised by 37.50 cm^3 of 0.20 mol dm^{-3} hydrochloric acid. What is the concentration of the sodium hydroxide?
NaOH (aq) + HCl (aq) → NaCl (aq) + H$_2$O (l)
Amount of HCl = 0.20×(37.50 / 1000) = 0.0075 mol
From the balanced equation, 1 mol of NaOH reacts with 1 mol of HCl.
Concentration of NaOH = 0.0075 / (25.00 / 1000) = 0.30 mol dm^{-3}

Example 3
In a titration analysis, 10.00 cm^3 of 0.40 mol dm^{-3} sulfuric acid is neutralised by 23.48 cm^3 of potassium hydroxide solution. What is the concentration of the KOH solution?
2KOH (aq) + H$_2$SO$_4$ (aq) → K$_2$SO$_4$ (aq) + 2H$_2$O (l)
Amount of H$_2$SO$_4$ = 0.40×(10.00 / 1000) = 0.0040 mol
From the balanced equation, 2 mol of KOH reacts with 1 mol of H$_2$SO$_4$
Amount of KOH = 2×0.0040 = 0.0080 mol
Concentration of KOH = 0.0080 / (23.48 / 1000) = 0.34 mol dm^{-3}

2.1 The nuclear atom
Syllabus
Nature of science – can you relate this topic to these concepts?
Evidence and improvements in instrumentation – alpha particles were used in the development of the nuclear model of the atom that was first proposed by Rutherford.
Paradigm shifts – the subatomic particle theory of matter represents a paradigm shift in science that occurred in the late 1800s.
Understandings – how well can you explain these statements?
Atoms contain a positively charged dense nucleus composed of protons and neutrons (nucleons).
Negatively charged electrons occupy the space outside the nucleus.
The mass spectrometer is used to determine the relative atomic mass of an element from its isotopic composition.
Applications and skills – how well can you do all of the following?
Use of the nuclear symbol notation to deduce the number of protons, neutrons and electrons in atoms and ions.
Calculations involving non-integer relative atomic masses and abundance of isotopes from given data, including mass spectra.

The basic structure of an atom
Atoms contain a positively charged dense nucleus composed of protons and neutrons (nucleons).
Negatively charged electrons occupy the space outside the nucleus.
This model has been developed over 200 years based on two strands of discovery:
- Discovery of electrons (Thomson), positively charged nucleus (Rutherford), charge on electrons (Millikan) and discovery of electrons (Chadwick).
- Postulated models of atoms: solid sphere (Dalton), plum pudding (Thomson) and the current model (based on Rutherford's Gold Foil experiment).

Timeline of discovery of sub-atomic particles

Year	Experiment	Scientist	Discovery
1897	Cathode rays	Thomson	Electrons are negatively charged
1909	Gold foil	Rutherford	Nucleus is positively charged
1910	Oil drop	Millikan	Charge on electron = 1.602×10^{-19} C
1932	Bombardment of Be with α-particles	Chadwick	Existence of the neutron

Models of the atom
Thomson's 'Plum Pudding' model (1807)
Negatively charged electrons (the plums) are embedded in a sphere of **uniform positive charge** (the pudding).

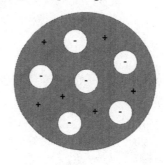

Dalton's Solid Sphere model (1808)
- All matter is made up of very small (indivisible) particles called **atoms**.
- An **element** consists of **one type of atoms only**.
- Different types of atoms **combine** in **whole-number ratios** to form **compounds**.
- In chemical reactions, atoms are **not created or destroyed**.

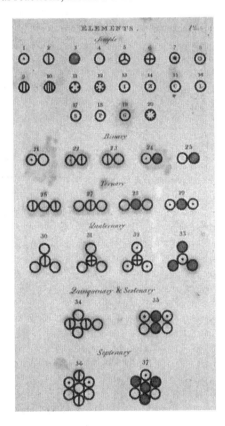

Rutherford's Planetary model (1911)
Rutherford's gold foil experiment

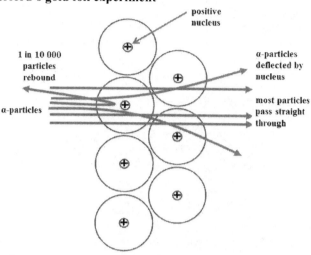

Conclusions:
- The electron cloud does not influence α-particle scattering.
- Much of an atom's **positive charge** is concentrated in a relatively tiny volume (the **nucleus**). The magnitude of this charge is proportional to the atomic mass. The nucleus is responsible for deflecting the α-particles.
- The nucleus of an atom of a particular element contains a fixed **number of protons (atomic number)**.
- The nucleus also contains **neutrons**. The number of neutrons varies from one **isotope** to another.

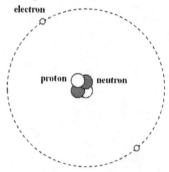

Size of atoms
Most atoms are between 1×10^{-10} to 5×10^{-10} m in diameter.
Others ways of saying this:
0.1 – 0.5 nm $1 \text{ nm} = 1\times10^{-9}$ m
100 – 500 pm $1 \text{ pm} = 1\times10^{-12}$ m
1 – 5 Å (angstrom) $1 \text{ Å} = 1\times10^{-10}$ m

1 unified atomic mass unit (amu) = 1.6605402×10^{-24} g = one-twelfth of the mass of a carbon-12 atom in its ground state (unexcited, stable state).

Particle	Mass (kg)	Mass (amu)	Charge (C)	Relative charge
Proton	1.673×10^{-27}	1	$+1.602\times10^{-19}$	+1
Neutron	1.675×10^{-27}	1	0	0
Electron	9.109×10^{-31}	1/1836 (negligible)	-1.602×10^{-19}	-1

Masses (definitions)
Relative atomic mass (A_r)
Relative Atomic Mass (A_r): ratio of the average mass of an atom to the unified atomic mass unit (amu).
A_r = average mass of an atom / unified atomic mass unit (amu)
The **average mass of an atom** is the **weighted average** of the atomic masses of all its **isotopes** and their **relative abundances**.
To determine relative atomic masses we need to measure relative isotopic masses and isotopic abundances. These data are obtained from using a **mass spectrometer**.
Relative molecular mass (M_r)
The average mass of a molecule divided by 1/12 the mass of a carbon-12 (^{12}C) atom.
Relative isotopic mass
The mass of an isotope of an element divided by 1/12 the mass of a carbon-12 (^{12}C) atom.

Isotopes
o   Isotopes of an element have the **same number of protons** but **different number of neutrons**.
o   The mass of an atom is almost entirely due to the protons and neutrons.
Number of electrons orbiting the nucleus = Number of

The nuclear symbol (isotopic notation)
The number of protons, neutrons and electrons can be deduced from the nuclear symbol for an atom (element).
X is the **symbol** for the element
Atomic number, Z = no. of protons in the nucleus
Mass number, A = no. of protons + no. of neutrons in the nucleus
In a neutral atom:
No. of protons = Z
No. of electrons = No. of protons
No. of neutrons = A - Z

Naming isotopes
Isotopes (nuclides) are specified by the name (or symbol) of the particular element by a hyphen and the mass number e.g. carbon-12, chlorine-37, uranium-235.
The chemical symbols are C-12, Cl-37, U-235 or generally

$$^{12}_{6}C, \ ^{37}_{17}Cl \ and \ ^{235}_{92}U$$

Since the atomic numbers are the same for all isotopes of the same element, it is often dropped from the notation i.e. ^{12}C, ^{37}Cl and ^{235}U.

Radioisotopes
Radioisotopes are isotopes which are radioactive – they are unstable and will decay into a more stable form by emitting radioactive radiation. The half life, $t_{½}$ is the time it takes for the amount of radioactive isotope to decrease to half of its initial value. This property makes them useful as:
- Medical tracers
- Medical treatments
- Archaeological dating

Uses of radioisotopes
Carbon-14 dating
Carbon-14 decays into nitrogen-14 when a neutron changes into a proton and an electron. The proton stays in the nucleus (which has now got an atomic number of 7) but the electron is ejected from the atom as a beta particle.

$$^{14}_{6}C \rightarrow \ ^{14}_{7}N + \ ^{0}_{-1}e$$

The relative abundance of carbon-14 in organisms remains constant as it is continually replenished by CO_2 in the atmosphere. When organisms die, no more carbon-14 is absorbed.
Carbon-14 decays with a half-life of 5730 years. The ratio of ^{14}C to ^{12}C can be used to determine the age of archaeological objects.

Cobalt-60 in radiotherapy
Cobalt-60 emits gamma radiation which is very penetrating. This can be used to destroy cancer cells.

The radiation damages the genetic material by knocking off electrons, making it impossible for the cell to grow or replicate.

Although normal cells are also damaged, they are able to recover with careful administration of treatment.

Iodine-131 as a medical tracer
Radioisotopes have the same chemical properties and play the same biological functions in the body. As they emit radiation, their positions can be traced in the body.

Iodine-131 emits beta and gamma radiation and has a half-life of 8 days.

Sodium iodide (NaI) is used to investigate the activity of the thyroid gland and to diagnose/treat thyroid cancer. The short half-life ensures that it is quickly eliminated from the body.

Iodine-125 pellets are implanted to the prostate gland to treat prostate cancer. It has a half-life of 80 days so that low levels of beta radiation can be emitted over an extended period.

Mass Spectrometer
To **determine relative atomic masses** we need to **measure relative isotopic masses** and **isotopic abundances**.
1. The sample is vaporised by an intensely hot filament.
2. The gaseous atoms are converted into positive ions by high energy electron bombardment (electron gun).
3. Beam of positive ions accelerated by a negatively charged electric field.
4. Positive ions are deflected by a perpendicular magnetic field according to their mass to charge (m/z) ratio. Lighter ions with higher charges are deflected more than heavier ions with lower charges.
5. A detector measures the relative abundance of each charged isotope present. As the positive ion hits the detector, an electrical pulse (current) is sent to a counter.

Sample applications
Mass spectrum of Zirconium
Isotopes present: ^{90}Zr, ^{91}Zr, ^{92}Zr, ^{94}Zr and ^{96}Zr.
Relative atomic mass =
$$\frac{(90 \times 51.5)+(91 \times 11.2)+(92 \times 17.1)+(94 \times 17.4)+(96 \times 2.8)}{100}$$
= **91.32**

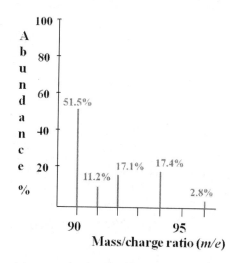

Mass spectrum of ethanol (C_2H_5OH)
The peak at m/e=46 is called the **molecular ion**.

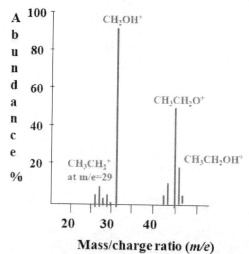

Why do elements have a non-integer A_r?
Isotopes – one element can have atoms with different masses e.g. carbon has 2 stable isotopes.

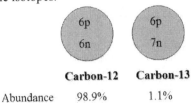

	Carbon-12	Carbon-13
Abundance	98.9%	1.1%

Why does chlorine have an A_r of approximately 35.5?

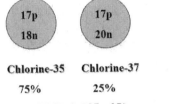

	Chlorine-35	Chlorine-37
Abundance	75%	25%

Calculation: $\dfrac{(35 \times 75.5) + (37 \times 25)}{100} = \underline{35.5}$

Notice each isotope has its own **relative isotopic mass** e.g. **chlorine-35 is 35, chlorine-37 is 37**.

In effect we are comparing the average relative isotopic masses of an element with that of the carbon-12 isotopic mass.

Calculation of relative atomic mass, A_r

The mass of an atom of an element divided by 1/12 the mass of a carbon-12 atom

The **general formula** used for calculating the relative atomic mass (A_r) of an element is:

$$A_r = \frac{\text{(relative isotopic mass × \% abundance of isotope-1)} + \text{(relative isotopic mass × \% abundance of isotope-2)} + \text{(relative isotopic mass × \% abundance of isotope-3)} + \text{etc}}{100}$$

Other uses of a Mass Spectrometer
- Determine the isotopic abundances and hence the A_r of elements.
- Detect drugs in urine samples. Most drugs e.g. steroids have unique molar masses.
- Identify elements from a mass spectrum profile.
- Carbon-14 dating: carbon-14 is radioactive and decays to nitrogen-14 with a half-life of 5730 years.
- Identify compounds and their molecular structures. Electron bombardment can also cause a molecule to break into smaller pieces. This is called **fragmentation.**

2.2 Electronic configuration

Nature of science – can you relate this topic to these concepts?
Developments in scientific research follow improvements in apparatus – the use of electricity and magnetism in Thomson's cathode rays.
Theories being superseded – quantum mechanics is among the most current models of the atom.
Use theories to explain natural phenomena – line spectra explained by the Bohr model of the atom.

Understandings – how well can you explain these statements?
Emission spectra are produced when photons are emitted from atoms as excited electrons return to a lower energy level.
The line emission spectrum of hydrogen provides evidence for the existence of electrons in discrete energy levels, which converge at higher energies.
The main energy level or shell is given an integer number, n, and can hold a maximum number of electrons, $2n^2$.
A more detailed model of the atom describes the division of the main energy level into s, p, d and f sub-levels of successively higher energies.
Sub-levels contain a fixed number of orbitals, regions of space where there is a high probability of finding an electron.
Each orbital has a defined energy state for a given electronic configuration and chemical environment and can hold two electrons of opposite spin.

Applications and skills – how well can you do all of the following?
Description of the relationship between colour, wavelength, frequency and energy across the electromagnetic spectrum.
Distinction between a continuous spectrum and a line spectrum.

Electromagnetic spectrum
- All electromagnetic waves travel at the **same speed**, which is $c = 3.00 \times 10^8$ ms^{-1}.
- Different types of waves have different wavelengths (λ) and frequencies (v).
- The **frequency** of a wave is the number of waves which pass a particular point in 1 s.
- f is inversely proportional to λ and related by: $c = v\lambda$

Different types of electromagnetic spectra

Continuous spectrum: consists of a continuous range of wavelengths e.g. dispersed visible light such as sunlight split by a prism.

Discrete spectrum: consists of distinct lines of specific frequencies.

Emission spectrum (discrete): a series of coloured lines on a dark background emitted by a gaseous element when an electrical discharge is applied.

Absorption spectrum (combination of discrete and continuous): a series of dark lines on a coloured background due to absorption of certain frequencies by a gaseous element. The background is a continuous spectrum broken up by a discrete spectrum of the absorbed frequencies.

Electron arrangement

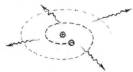

The main problem with Rutherford's model of the atom: Why doesn't the orbiting electron spiral towards the nucleus as it loses energy?

Energy in discrete parcels (quantization) – Max Planck

Electromagnetic radiation has been traditionally described in terms of its **wave** properties i.e. wavelength and frequency. With the advent of Quantum Mechanics in the 20th century, there has been a focus on the particle-like properties of electromagnetic radiation.

Electromagnetic radiation is transmitted in discrete parcels of energy called **photon**. The **energy (E)** of the photon is related to the **wavelength (λ)** and **frequency (ν)** of the radiation.

$$E = h\nu = \frac{hc}{\lambda}$$

h = Planck's constant = 6.63×10^{-34} Js
ν = frequency of radiation in Hz
c = speed of light = 3.00×10^8 ms^{-1}
λ = wavelength of radiation in m

The energy of an electron is **quantised** (it can only have certain discrete levels of energy).

Energy Levels and the Line Spectra (of hydrogen) – Niels Bohr

- Elements in the gaseous state can be **excited** by heating or being placed in a discharge tube. **Spectral lines** of different frequencies are emitted with some colours of the continuous spectrum missing.
- The electron in hydrogen can only exist at discrete **circular orbits** around the nucleus called **quantum shells**. The energy (E) of this electron is fixed or **quantized**.

$$E = -R_H \left(\frac{1}{n^2}\right)$$

R_H = Rydberg constant = 2.18×10^{-18} J
n = principal quantum number (energy level) = 1, 2, 3, 4, …

Emission spectra

- When hydrogen atoms are heated, electrons are **promoted** from the '**ground state**' to a higher, '**excited**' state.
- The electron then drops back to the ground state, giving out the energy as a **photon** of **light**. The energy of the photon is equal to the energy change in the atom and related to the frequency of radiation by Planck's constant.

$\Delta E_{electron} = E_{photon} = hf$ h = Planck's constant

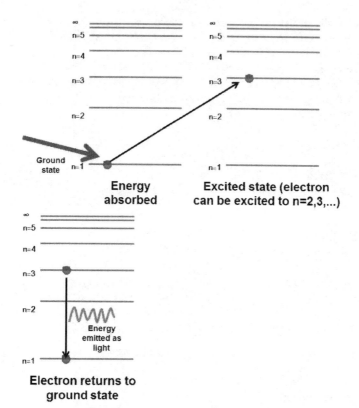

- **Lyman series**: electron drops from an excited state back to the ground state (n=1).
- **Balmer series**: electron drops from an excited state back to the second shell (n=2).
- **Paschen series**: electron drops from an excited state back to the second shell (n=3).

The energy of the photon, ΔE is equal to the energy change in the atom and related to the frequency of radiation by Planck's constant, h.

$$\Delta E = E_f - E_i = h\upsilon = \frac{hc}{\lambda}$$

E_f = energy of the electron in the final state
E_i = energy of the electron in the initial state

$$\Delta E = \left[-R_H\left(\frac{1}{n_f^2}\right)\right] - \left[-R_H\left(\frac{1}{n_i^2}\right)\right] = \left[R_H\left(\frac{1}{n_i^2} - \frac{1}{n_f^2}\right)\right]$$

If electrons are excited beyond n=∞ (the continuum), it no longer experiences an attraction to the nucleus. The electron is said to be a **free electron** and the atom is **ionised** (acquires a **positive charge** due to an excess of protons.

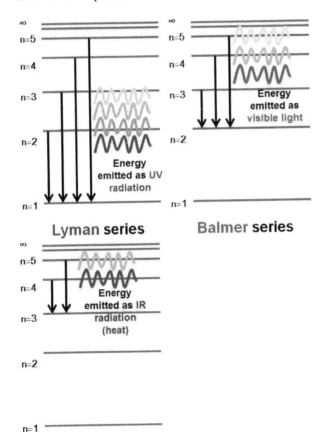

- The energy levels (shells) in an atom can only hold a finite number of electrons.
- The first shell can hold up to a maximum of 2 electrons.
- The second and third shells can hold up to a maximum of 8 electrons before the fourth shell starts getting filled.

Ionisation energy (IE or ΔH_i)

First ionisation energy: The energy required to remove **one electron** from the **ground state** of each **gaseous atom** in a **mole** of an element.

$A\,(g) \rightarrow A^+\,(g) + e^-$ ΔH_{i1} (positive) kJ mol^{-1}

- Ionisation is an **endothermic** process.
- It is the energy required to move the electron from the n=1 level to n=∞ (beyond the influence of the nucleus).

Second ionisation energy: The energy required to remove **one electron** from each **gaseous singly charged ion** in a **mole** of an element.

$A^+(g) \rightarrow A^{2+}\,(g) + e^-$ ΔH_{i2} (positive) kJ mol^{-1}

- The **second ionisation energy is higher** because:
- The electron is removed from a positive ion which is smaller, and the electron is closer to the nucleus.
- The electron is more attracted to the nucleus (more protons than electrons).

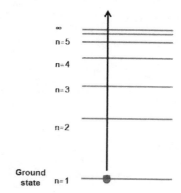

Sublevels and orbitals

n=1
The first shell (n=1) is not divided any further.
It is spherical and designated by 1s.

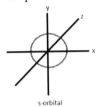

s-orbital

n=2
The second shell (n=2) is divided into two types of orbitals.
The 2s orbital is spherical but larger than 1s.
The three 2p orbitals are dumbbell shaped and point along the three axes x, y and z. They have the same energies but slightly higher than 2s.

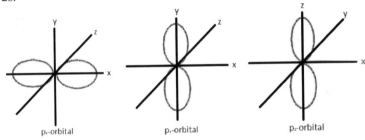

p_x-orbital p_y-orbital p_z-orbital

n=3
The third shell (n=3) consists of three subshells with different types of orbitals.
The 3s and 3p orbitals are similar to 2s and 2p but bigger.
The third subshell has five 3d orbitals.

n=4
The fourth shell (n=4) contains one 4s, three 4p, five 4d and seven 4f orbitals.

Heisenberg's Uncertainty Principle

It is impossible to make an exact and simultaneous measurement of both the **position** and **momentum (movement)** of any given body. Applied to electrons, it is impossible to know the exact position of an electron at any instant. The **orbitals** show the region of space (locus) of where there is a probability of finding an electron and this can be calculated using **Schrödinger's equation**.

Schrödinger's equation

Probability density, ψ^2 = probability of finding an electron in a region of space at a point, r from the nucleus.

ψ = wavefunction (describes the possible energy states of the electron in a hydrogen atom.

Electron spin and the spin magnetic quantum number, m_s

Each orbital can contain a maximum of **two electrons**. The electrons are said to be 'spinning' and this generates a **magnetic field** around them. The **orientation** of the fields are described by the **spin magnetic quantum number, m_s** which can have a value of $+\frac{1}{2}$ or $-\frac{1}{2}$.

When two electrons are in the same orbital, their spins must be **opposite** i.e. the magnetic fields cancel each other out.

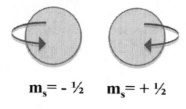

$m_s = -\frac{1}{2}$ $m_s = +\frac{1}{2}$

Order of energies for the orbitals

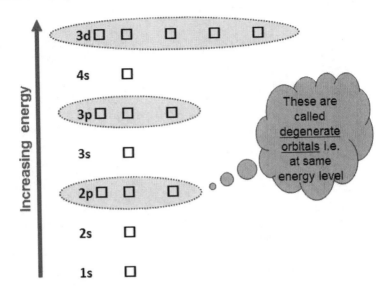

Filling orbitals (the *aufbau* method)
1. Electrons occupy orbitals with the <u>lowest energy</u>.
2. **Pauli exclusion principle**:
   An orbital can contain a <u>maximum</u> of <u>2 electrons</u>.
   If there are 2 electron in the same orbital, they are of <u>opposite spins</u>.
3. **Hund's rule**: Electrons will fill a set of degenerate orbitals by keeping their spins parallel.

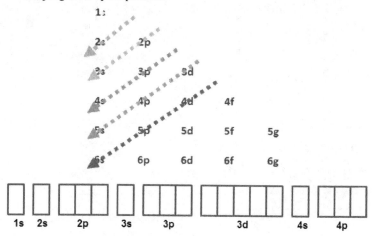

Principal quantum number, n	Subshells	No. of orbitals in subshell	Max. no. of electrons in subshells	Max. no. of electrons in energy level (shell)
1	s	1	2	2
2	s	1	2	8
	p	3	6	
3	s	1	2	18
	p	3	6	
	d	5	10	
4	s	1	2	32
	p	3	6	
	d	5	10	
	f	7	14	

Unabbreviated electron configurations
sodium : $1s^22s^22p^63s^1$
iron: $1s^22s^22p^63s^23p^63d^64s^2$
bromine: $1s^22s^22p^63s^23p^63d^{10}4s^24p^5$
calcium: $1s^22s^22p^63s^23p^64s^2$
chromium: $1s^22s^22p^63s^23p^63d^54s^1$
Abbreviated electron configurations
cobalt: [Ar] $4s^23d^7$
silver: [Kr] $5s^14d^{10}$
manganese: [Ar] $3d^54s^2$
copper: [Ar] $3d^{10}4s^1$
zinc: [Ar] $3d^{10}4s^2$
More examples
$1s^22s^22p^63s^23p^4$: **sulphur**
$1s^22s^22p^63s^23p^63d^{10}4s^24p^1$: **gallium**
[Ar] $3d^14s^2$: **scandium**
[Ne] $3s^23p^63d^34s^2$: **vanadium**
[Ar] $4s^1$: **potassium**
Validity of electronic configurations
$1s^22s^22p^63s^23p^64s^24d^1$: **not valid (take a look at "4d")**
$1s^22s^22p^63s^33p^53d^54s^2$: **not valid (3p must be filled)**
[Se] $4s^24p^3$: **not valid (Se isn't a noble gas)**
[Ne] $3d^{10}4s^2$: **not valid (3s and 3p must be shown**
[Ar] **not valid (need to be written out)**

12.1 Electrons in atoms
Syllabus
Nature of science – can you relate this topic to these concepts?
Experimental evidence to support theories—emission spectra provide evidence for the existence of energy levels.
Understandings – how well can you explain these statements?
In an emission spectrum, the limit of convergence at higher frequency corresponds to the first ionization energy.
Trends in first ionization energy across periods account for the existence of main energy levels and sub-levels in atoms.
Successive ionization energy data for an element give information that shows relations to electron configurations.
Applications and skills – how well can you do all of the following?
Solving problems using $E = h\nu$.
Calculation of the value of the first ionization energy from spectral data which gives the wavelength or frequency of the convergence limit.
Deduction of the group of an element from its successive ionization energy data.
Explanation of the trends and discontinuities in first ionization energy across a period.

Wave-particle duality (de Broglie equation)
The **de Broglie equation** links the wave and particle-like properties of the electron.

$$\lambda = \frac{h}{p}$$

λ = wavelength (a wave property)
h = Planck's constant
p = momentum = m x v (a particle property)
Implication: very small particles e.g. electrons which are moving will exhibit wave-like properties as the wavelength will be sufficiently large enough to observe.

Heisenberg's uncertainty principle (extension)

$$\Delta p \times \Delta q \geq \frac{h}{4\pi}$$

Δp = uncertainty in momentum measurement
Δq = uncertainty in position measurement
$\Delta p \rightarrow 0, \Delta q \rightarrow \infty$ and $\Delta p \rightarrow 0, \Delta q \rightarrow \infty$
Implication: a very small uncertainty (by measuring accurately) in one of the properties results in an infinitely large uncertainty in the other measurement.

Ionisation energy (IE)
First ionisation energy (IE$_1$): The energy required to remove **one electron** from the **ground state** of each **gaseous atom** in a **mole** of an element.
$A(g) \rightarrow A^+(g) + e^-$ IE$_1$ (positive) kJ mol^{-1}
Ionisation is an endothermic process. It is the energy required to move the electron from the n=1 level to n=∞ (beyond the influence of the nucleus).
Second ionisation energy (IE$_2$): The energy required to remove **one electron** from each **gaseous singly charged ion** in a **mole** of an element.
$A^+(g) \rightarrow A^{2+}(g) + e^-$ IE$_2$ (positive) kJ mol^{-1}
The second ionisation energy is higher because:
- The electron is **more attracted to the nucleus** (more protons than electrons) – major factor.
- The electron is removed from a positive ion which is smaller, and the electron is **closer to the nucleus**.

Successive ionisation energies will have the positive values increase in succession due to the reasons stated above.

Successive ionisation energy

Successive ionisation energies for aluminium (in kJ mol^{-1}):

1st	2nd	3rd
580	1820	2750

4th	5th	6th	7th	8th	9th	10th	11th
11600	14800	18400	23300	27500	31900	38500	42700

12th	13th
200000	222000

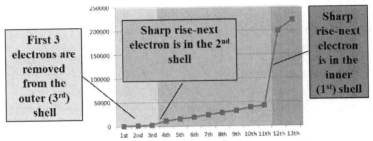

Successive ionisation energies can be used to:
o provide evidence for the existence of electron shells.
o deduce the group number of the element.

The data below show the first 9 successive ionisation energies of an element in units of kJmol^{-1}.
From which group in the Periodic Table does the element come?

590, 1145, 4912, 6474, 8144, 10496, 12320, 14207, 18192
Group 2

789, 1577, 3232, 4356, 16091, 19785, 23787, 29253, 33878
Group 4

496, 4563, 6913, 9544, 13352, 16611, 20115, 25491, 28934
Group 1

1000, 2251, 3361, 4564, 7012, 8496, 27107, 31671, 36579
Group 6

3.1 Periodic Table
Syllabus
Nature of science – can you relate this topic to these concepts?
Obtain evidence for scientific theories by making and testing predictions based on the – scientists organize subjects based on structure and function; the periodic table is a key example of this. Early models of the periodic table from Mendeleev, and later Moseley, allowed for the prediction of properties of elements that had not yet been discovered.

Understandings – how well can you explain these statements?
The periodic table is arranged into four blocks associated with the four sub-levels – s, p, d, and f.
The periodic table consists of groups (vertical columns) and periods (horizontal rows).
The period number (n) is the outer energy level that is occupied by electrons.
The number of the principal energy level and the number of the valence electrons in an atom can be deduced from its position on the periodic table.
The periodic table shows the positions of metals, non-metals and metalloids.

Applications and skills – how well can you do all of the following?
Deduction of the electron configuration of an atom from the element's position
on the periodic table, and vice versa.

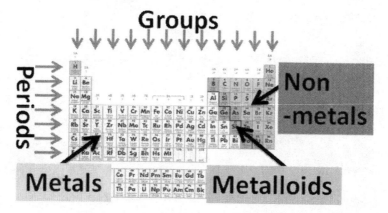

Historical development – Döbereiner (Law of Triads)
Döbereiner (1817) noticed a pattern when considering the A_r of groups of 3 elements:

Element	Li		Na		K
A_r	7	+16	23	+16	39
		½ (7+39) = 22 ≈ 23			

Element	S		Se		Te
A_r	32	+47	79	+49	128
		½ (32+128) = 80 ≈ 79			

Element	Cl		Br		I
A_r	35.5	+44.5	80	+47	127
		½ (35.5+127) = 81.3 ≈ 80			

Element	Ca		Sr		Ba
A_r	40	+48	88	+49	137
		½ (40+137) = 88.5 ≈ 88			

Development of the Periodic Table
Historically, elements were recognised as showing properties which could be arranged in many ways.
Properties which were considered were:
- physical state.
- metal/non-metal.
- reactions with water, oxygen etc.
- relative atomic mass.

Historical development – Newlands (Law of Octaves)
Newlands (1865) noticed a repetitive pattern when considering the increasing A_r values of elements. They appeared to behave in eights. Unfortunately, he could only apply this to the first 16 elements.

Historical development – Mendeleev (Periodic Law)
Mendeleev (1869) produced the most comprehensive organisation of the known elements – the Periodic Table. The elements were ordered in increasing relative atomic mass. He noticed a regular repeating pattern of properties. To achieve this he made some crucial decisions:
- He left spaces to allow similar elements to appear in the same group.
- He predicted the spaces would be filled by elements yet to be discovered.
- The properties of these undiscovered elements could be predicted.

Historical development – Moseley's modifications
Moseley (1913) proposed the real property responsible for periodicity was atomic number, Z i.e. the number of protons in the nuclei of atoms.

Metals
Physical properties
- **Conduct** electricity and heat (in solid and molten form).
- **Malleable** – can be shaped by applying a force.
- **Ductile** – can be drawn into long, thin wires.
- **Solids** at **room temperature** (except mercury).
- **Shiny** (have **lustre**).

Chemical properties
- Form **cations** (positive ions) e.g. Na^+, Mg^{2+}
- **Reactivity increases down a group** i.e. reactivity: K > Na > Li
  If reactive enough, react with:
  1. Acids to form a salt and hydrogen gas e.g.
     magnesium + hydrochloric acid → magnesium chloride + hydrogen
  2. Water to form a hydroxide and hydrogen gas e.g.
     sodium + water → sodium hydroxide + hydrogen
  3. Displaces less reactive cations of metals from their compounds.
     copper + silver nitrate → copper nitrate + silver
- **Oxides** and **hydroxides** are **bases**.

Non-metals
- Form **anions** (negatively charged ions) in many compounds with metals.
- Form **covalent bonds** with non-metals.
- **Reactivity increases up a group** i.e. reactivity: F > Cl > Br > I
- **Oxides** are **acidic** (react or dissolve in water to form an acidic solution).

Metalloids
- Have metallic and non-metallic properties.
- Do not form ions.
- **Semiconductors** of electricity e.g. Si and Ge.
- Oxides are **weakly acidic**.

Valence electrons (outer-shell electrons)
For s-block elements, no. of valence electrons = group no.
For p-block elements, no. of valence electrons = group no. - 10

The Modern Periodic Table
The modern version shows **vertical groups** and **horizontal periods** displayed in blocks.
- **s-block**: Groups 1 and 2.
- **p-block**: Groups 3 to 7 and Group 0.
- **d-block**: Transition elements.
- **f-block**: Lanthanide and Actinide element.

Physical properties
Many different physical and chemical properties of elements fit into periodic patterns in the table:
- Structure & bonding.
- Electronic configuration and valency.
- Metallic properties.
- Atomic and ionic radii (size of atoms or ions).
- Melting and boiling points.
- Trends emerging across the table from elements 1 to 36 (H to Kr).
    - Giant metallic lattice: Groups 1-13 and Transition Metals.
    - Giant molecular lattice: Groups 13-15.
    - Molecules: Groups 15-17 and hydrogen.
    - Atoms: Group 0.

Electronic configurations and valency
- Elements in the same group have the same outer shell arrangement of electrons.
  Mg: $1s^2 2s^2 2p^6 \mathbf{3s^2}$
  Ca: $1s^2 2s^2 2p^6 3s^2 3p^6 \mathbf{4s^2}$
- In the **s-block**, the outer-most electrons are in the **s-orbital**. They usually lose their outer-most electron to form positive ions.
- In the **p-block**, the outer-most electrons are in the **p-orbitals**.
- Group 0 elements have **full outer shells of electrons**.

Metallic properties
Metallic properties **decrease from left to right** (across a period).
Metallic properties **increase down a group**.
Metals are **malleable** (can be hammered/pressed) and **ductile** (can be drawn out into wires).
Metals have a **higher electrical conductivity** than molecular elements.
Metals form **positive ions (cations)** in compounds by losing one or more electrons.
Metals react with oxygen to form oxides that are **bases** (substance that accepts H^+ ions from an acid).
Metals that are reactive react with acids to form a salt and hydrogen gas.

3.2 Periodic trends
Syllabus
Nature of science – can you relate this topic to these concepts?
Looking for patterns – the position of an element in the periodic table allows scientists to make accurate predictions of its physical and chemical properties. This gives scientists the ability to synthesize new substances based on the expected reactivity of elements.
Understandings – how well can you explain these statements?
Vertical and horizontal trends in the periodic table exist for atomic radius, ionic radius, ionization energy, electron affinity and electronegativity.
Trends in metallic and non-metallic behaviour are due to the trends above.
Oxides change from basic through amphoteric to acidic across a period.
Applications and skills – how well can you do all of the following?
Prediction and explanation of the metallic and non-metallic behaviour of an element based on its position in the periodic table.
Discussion of the similarities and differences in the properties of elements in the same group, with reference to alkali metals (group 1) and halogens (group 17).
Construction of equations to explain the pH changes for reactions of Na_2O, MgO, P_4O_{10}, and the oxides of nitrogen and sulfur with water.

Elemental structure

Group no.	1 through 12
Structure	giant metallic lattice
Element Type	light metals, transition metals, heavy metals

Group no.	13	14	15	16	17	18
Structure	giant molecular lattice → simple molecular					Atoms
Element Type	metalloids, non-metals					noble gases

Forces on an electron in an atom
Electrons are attracted to the nucleus by **electromagnetic forces**.
The size of the force depends on **Coulomb's law**:

$$\text{force} \propto \frac{q_+ q_-}{r^2}$$

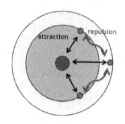

q_+ : charge of positive object (nucleus)
q_- : charge of negative object (electron)
r : distance between the centres of the objects

- The **greater the atomic number** (no. of protons), the **stronger the attraction** on the electrons.
- Electrons that are **further away** from the nucleus are **less attracted** towards it.
- **Electrons repel each other** – inner electrons repel outer electrons more because they are more densely packed.

Shielding effect and effective nuclear charge
The **core (non-valence) electrons 'shield'** the outer-shell electrons from the attractive forces of the nucleus.
Effective nuclear charge (Z_{eff}): Net charge on an electron, after allowing for the electrons in orbit around the nucleus shielding the full (positive) nuclear charge.
$Z_{eff} = Z - S$
Z = actual nuclear charge (atomic number)
S = screening / shielding constant
The actual calculations using this equation (Slater's rules) are very complicated.
For IB Chemistry, it is sufficient to imagine that S approximately equals the number of core electrons.

Across Period 1
Hydrogen – $Z = 1$; No. of electrons = 1 ($1s^1$)
$Z_{eff} = 1$ (the electron feels the full force)
Helium – Nuclear charge = 2; No. of electrons = 2 ($1s^2$)
$Z_{eff} \approx 2$ (the two electrons repel each other, the net attraction between the nucleus and each electron is less)

Across Period 2
Lithium – $Z = 3$; No. of electrons = 3 ($1s^2\ 2s^1$); $S \approx 2$
$Z_{eff} \approx 3 - 2 \approx 1$ (the 2s electron is shielded by the two core 1s electrons)
Beryllium – $Z = 4$; No. of electrons = 4 ($1s^2\ 2s^2$); $S \approx 2$
$Z_{eff} \approx 4 - 2 \approx 2$ (the 2s electrons are shielded by the two core 1s electrons, the two 2s electrons repel each other slightly)
Boron – $Z = 5$; No. of electrons = 5 ($1s^2\ 2s^2\ 2p^1$); $S \approx 2$
$Z_{eff} \approx 5 - 2 \approx 3$ (the 2s and 2p electrons are shielded by the two core 1s electrons, the two 2s electrons repel the 2p electron as they are closer to the nucleus)

Down a group
Lithium – $Z = 3$; No. of electrons = 3 ($1s^2\ 2s^1$); $S \approx 2$
$Z_{eff} \approx 3 - 2 \approx 1$ (the 2s electron is shielded by the two core 1s electrons)
Sodium – $Z = 11$; No. of electrons = 11 ($1s^2\ 2s^2\ 2p^6\ 3s^1$); $S \approx 10$
$Z_{eff} \approx 11 - 10 \approx 1$ (the 3s electron is shielded by the core 1s, 2s and 2p electrons)
Potassium – $Z = 19$; No. of electrons = 19 ($1s^2\ 2s^2\ 2p^6\ 3s^2\ 3p^6\ 4s^1$); $S \approx 18$
$Z_{eff} \approx 19 - 18 \approx 1$ (the 4s electron is shielded by the core 1s, 2s, 2p, 3s, and 3p electrons)

The effective nuclear charge is approximately the same for all members of a group in the periodic table.

Atomic and Ionic Radii
Bonding atomic radius (covalent radius), R_b

½ x distance between the nuclei of two non-metallic atoms chemically bonded together

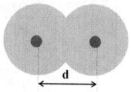

$R_b = ½ \times d$

The bonding atomic radius depends on the strength of the bond and the interactions between the nuclei and the electrons. This varies between different compounds. Therefore, the values for atomic radii given in the Data Booklet are taken from **experimental mean values** from a wide range of elements and compounds.

Non-bonding atomic radius (van der Waals' radius), R_{nb}

½ x distance between the nuclei of two atoms which are not chemically bonded e.g. noble gases

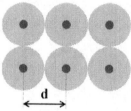

$R_{nb} = ½ \times d$

Atomic Radii trends

A plot of atomic (covalent) radii against atomic number for elements 1 to 35 shows the following trends:

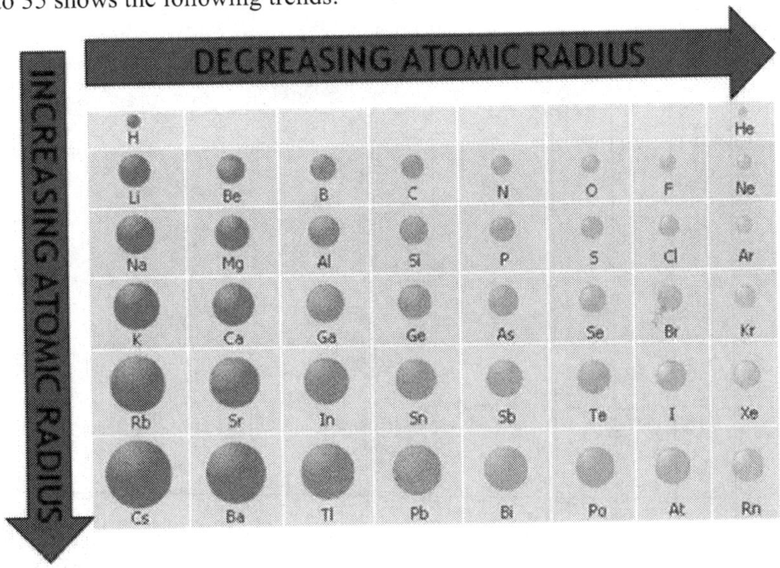

Atomic radii

- **increase down a Group** - the number of occupied electron shells increase. The shielding effect of the core electrons increases with the number of core shells.
- **decrease across a Period** - the greater nuclear charge increases the force of attraction on the electrons, drawing them closer to the nucleus. The effective nuclear charge, Z_{eff} increases.
- are **relatively constant** across the transition elements Ti to Cu after an initial decrease.

Ionic Radii trends
A positive ion is always smaller than its neutral atom because:
o   the number of electrons decrease.
o   fewer electrons mean less repulsion between these electrons

If the ions have the same electronic configuration (isoelectronic), the ion with the greatest charge will have the smallest ionic radius e.g. $Al^{3+} < Mg^{2+} < Na^+$

A negative ion is always larger than its neutral atom because:
o   the number of electrons increase.
o   more electrons mean more repulsion between these electrons, causing the ion to expand.

For isoelectronic ions (ions with the same number of electrons):
o   Groups 1 to 14 positive ions (Period 3)
    Na^+, Mg^{2+}, Al^{3+}, Si^{4+} have the same electron arrangement (2,8). Ionic radius decreases across the period due to the increasing atomic number (nuclear charge) which increases the attraction between the nucleus and the electrons.
o   Groups 14 to 17 negative electrons (Period 3)
    Si^{4-}, P^{3-}, S^{2-}, Cl^- have the same electron arrangement (2,8,8). Ionic radius decreases across the period (same reason).

The positive ions are smaller than the negative ions as they have one occupied electron shell.

The ionic radii increase down the group as the number of occupied electron shells increases.

Melting and boiling point trends

A plot of boiling point of elements 1 to 36 against atomic number reveals a pattern with 'peak' values for elements in Group 14 i.e. carbon and silicon.

Carbon (graphite) has a very high melting point due to the **strong covalent bonds** between the atoms.

Silicon has a very high melting point due to the strong covalent bonds and formation of a **giant molecular lattice**.

Across periods 2 and 3

The melting point increases steadily for the first four elements and then fall sharply at the fifth.

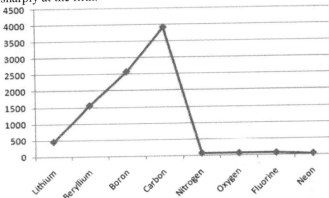

The trend is similar for boiling points.
The first three elements are metals.

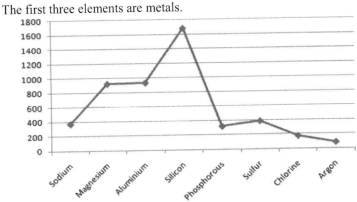

The peak at the fourth element is due to giant covalent structure.
The sharp fall is because the last four elements have simple covalent or single atomic structures.

Down Groups 1 and 2 (metals)

There is a general decrease in melting point. The structure is held together by the attraction between delocalised outer electrons and the positively charged ions.

As the number of delocalised electron per atom and the charge on the ion is the same in a Group, the force of attraction decreases with increasing atomic radius.

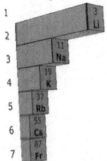 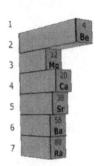

The trend is similar for boiling points.

Down Group 17

There is a general increase in the melting point. The elements exist as diatomic molecules held together by van der Waals' intermolecular forces. These forces increase in strength with the number of electrons in the molecule.

At room temperature:
- F_2 and Cl_2 are gases.
- Br_2 is a liquid.
- I_2 and At_2 are solids.

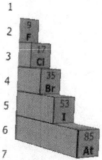

The trend is similar for boiling points.

Electronegativity

Electronegativity (χ_P): The extent to which an element attracts a pair of electrons in a covalent bond towards itself. This is measured on the Pauling scale.
- **Increases across a period** – increasing nuclear charge (bonding electrons are increasingly attracted to the nucleus).
- **Decreases down a group** – increasing number of electron shells and radius (bonding electrons are further away from the nucleus). The increasing nuclear charge is shielded by the core electrons.

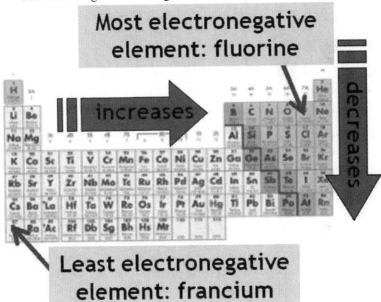

First ionisation energy trends

First ionisation energy (IE$_1$): The energy required to remove **one electron** from the **ground state** of each **gaseous atom** in a **mole** of an element.

$A (g) \rightarrow A^+ (g) + e^-$ IE$_1$ (positive) kJ mol^{-1}

Second ionisation energy (IE$_2$): The energy required to remove **one electron** from each **gaseous singly charged ion** in a **mole** of an element.

$A^+(g) \rightarrow A^{2+} (g) + e^-$ IE$_2$ (positive) kJ mol^{-1}

Factors influencing ionisation energy values are:
1. Increase in positive nuclear charge will cause an increase in first ionisation energy – electrons experience more attraction to the nucleus.
2. Forces of attraction between positive nuclear charge and negatively charged electrons decreases as the number of shells in an atom increases – electrons are further away from the nucleus.
3. Completely filled inner electron shells shield outer electrons from the nucleus and repel them – effective nuclear charge on the outer electrons decrease.

Down a group

ΔH$_{i1}$ decreases – the electron removed is from an electron shell further from the nucleus. Although the nuclear charge increases, the effective nuclear charge is about the same due to shielding by the inner electrons.
Li>Na>K; Be>Mg>Ca

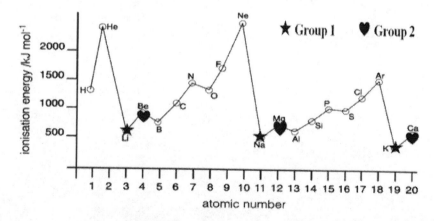

Across a period
ΔH_{i1} **generally increases** (effective nuclear charge increases). The variations are evidence of subshells.
Be→B: The 2p electron is at a higher energy level than the 2s electrons, and also partially shielded by the 2s electrons. Less energy is needed to remove a 2p electron.
N→O: N ($1s^2\ 2s^2\ 2p_x^1\ 2p_y^1\ 2p_z^1$) and O ($1s^2\ 2s^2\ 2p_x^2\ 2p_y^1\ 2p_z^1$). The two paired electrons in the $2p_x$ orbital of oxygen repel each other (easier to remove).
The explanations are similar for Mg→Al and P→S.

First Ionisation Energies and reactivity
Ionisation energies give a measure of how easily elements change to positive ions.
$M(g) \rightarrow M^+(g) + e^-$
Formation of positive ions is important in many reactions e.g. metals with water, oxygen, chlorine.
$Na(s) \rightarrow Na^+(aq) + e^-$
Group 1-13 elements (metals) with low ionisation energies will react more quickly and vigorously.

Electron affinity (E_{ea})
Electron affinity **generally becomes more negative across a Period**. Exceptions occur when:
- adding a first electron into an empty p sublevel (E_{ea} less negative).
- adding a second electron into the p sublevel – this electron experiences extra repulsion from the other electron (E_{ea} less negative than expected).
- adding a third electron into the p sublevel – the p sublevel is half-filled with electrons with parallel spin (E_{ea} more negative).

The next three electrons to fill the p sublevel show a similar trend.

Period 4 element	K	Ca	Ga	Ge	As	Se	Br	Kr
E_{ea} / kJ mol^{-1}	-48	-2	-41	-119	-78	-195	-325	>0

Electron affinity **generally becomes less negative down a Group**. This is because the attraction on the incoming electron becomes weaker as the effective nuclear charge is shielded by an increasing number of inner shells.

Group 1 element	Li	Na	K	Rb	Cs	Fr
E_{ea} / kJ mol^{-1}	-60	-53	-48	-47	-46	-47

Metallic properties

Metallic properties **decrease from left to right** (across a period).
Metallic properties **increase down a group**.
Metals tend to **lose electrons** to form **positive ions (cations)** – they are **oxidized**.
$M(g) \rightarrow M^+(g) + e^-$
Formation of positive ions is important in many reactions e.g. metals with water, oxygen, chlorine. Metals with **low ionisation energies** will react more quickly and vigorously.
Non-metals tend to **accept electrons** to form **negative ions (anions)** – they are **reduced**.
$X(g) + e^- \rightarrow X^-(g)$

Elements which are **more electronegative** usually have highly **negative electron affinities**.

	1	2	13	14	15	16	17
1							
2	Li^+				N^{3-}	O^{2-}	F^-
3	Na^+	Mg^{2+}	Al^{3+}		P^{3-}	S^{2-}	Cl^-
4	K^+	Ca^{2+}				Se^{2-}	Br^-
5	Rb^+	Sr^{2+}				Te^{2-}	I^-
6	Cs^+	Ba^{2+}					

Chemical properties
Group 0: the noble gases
Helium (He) - Electronic arrangement: 2
Neon (Ne) - Electronic arrangement: 2,8
Argon (Ar) - Electronic arrangement: 2,8,8
Krypton (Kr) - Electronic arrangement: 2,8,18,8
Xenon (Xe) - Electronic arrangement: 2,8,18,18,8
Radon (Rn) - Electronic arrangement: 2,8,18,32,18,8

All the elements in this group:
- have **fully occupied outer shells**.
- are **monatomic colourless gases** – they exist as individual atoms.
- very chemically **unreactive**
    - all the elements have a full outer shell of electrons (helium has two, all the other noble gases have eight) - they do not form chemical bonds with each other or other elements.
    - removing an outer electron requires a large amount of energy.
    - gaining an electron involves adding it to an empty outer shell which experiences negligible effective nuclear force – the protons in the nucleus are shielded by an equal amount of inner shell electrons.

Group 1: the alkali metals
All the elements in this group:
- have **low densities** compared to other metals - lithium, sodium and potassium are all less dense than water and so will float.
- have **low melting points** compared to other metals.
- are good **conductors of electricity** due to the mobile outer electron.
- have shiny grey surfaces when freshly cut with a knife.

Element	Symbol	Density (g cm^{-3})	Melting point (°C)
Lithium	Li	0.53	181
Sodium	Na	0.97	98
Potassium	K	0.86	63
Rubidium	Rb	1.53	39
Caesium	Cs	1.88	29

All the elements in this group have 1 electron in the outermost shell and are **very reactive**.
- Reactions involve the **loss of the valence electron** which changes the metal atom into a **M⁺ ion**.
- Reactivity **increases down the group** as the atomic number and radius increases. The **ionisation energies decrease down** the group because:
    - The valence electron gets **further from the nucleus** and reduces the force of attraction.
    - The inner shells **'shield'** the valence electron from the attraction of the nucleus.

Alkali metal	Observation and equations
Lithium	Reacts **slowly**. **Floats** on water. **Effervescense (bubbling)** observed as hydrogen gas is released. $2Li\ (s) + 2H_2O\ (l) \rightarrow 2LiOH\ (aq) + H_2\ (g)$
Sodium	Reacts **vigorously**. The metal **melts** due to the heat generated and **forms a ball** (takes up the least amount of space). **Effervescense (bubbling)** observed as hydrogen gas is released. $2Na\ (s) + 2H_2O\ (l) \rightarrow 2NaOH\ (aq) + H_2\ (g)$
Potassium	Reacts **violently**. **Effervescense (bubbling)** observed as hydrogen gas is released which **ignites** instantly to burn with a **lilac-coloured flame** due to the intense heat generated. $2K\ (s) + 2H_2O\ (l) \rightarrow 2KOH\ (aq) + H_2\ (g)$
Rubidium	**Explodes** on contact with water. $2Rb\ (s) + 2H_2O\ (l) \rightarrow 2RbOH\ (aq) + H_2\ (g)$
Caesium	**Explodes** on contact with water. $2Cs\ (s) + 2H_2O\ (l) \rightarrow 2CsOH\ (aq) + H_2\ (g)$

Group 17: the halogens
- highly reactive non-metals
- exist as diatomic molecules containing a covalent bond
- due to their reactivity the elements do not occur in the free state (always combined with other elements or itself)

Element	Appearance at room temperature
Fluorine (F_2)	Pale yellow gas
Chlorine (Cl_2)	Greenish yellow gas
Bromine (Br_2)	Brown liquid, gives off red vapour
Iodine (I_2)	Dark grey solid, sublimes to a purple vapour

Element	Melting point (K)	Boiling point (K)
Fluorine (F_2)	53.5	85.0
Chlorine (Cl_2)	172	239
Bromine (Br_2)	266	332
Iodine (I_2)	387	457

Element	Electronic configuration
Fluorine (F_2)	[He] $2s^2 2p^5$
Chlorine (Cl_2)	[Ne] $3s^2 3p^5$
Bromine (Br_2)	[Ar] $3d^{10} 4s^2 4p^5$
Iodine (I_2)	[Kr] $4d^{10} 5s^2 5p^5$

Element	Atomic radius (nm)	Ionic radius (nm)
Fluorine (F_2)	0.7	0.14
Chlorine (Cl_2)	0.10	0.18
Bromine (Br_2)	0.11	0.20
Iodine (I_2)	0.13	0.22

All the elements in this group have **7 valence electrons** and are **very reactive**.

Reactions involve the **gain of an electron** which changes the halogen atom into a **halide, X^- ion**.

Reactivity **decreases down the group** as the atomic number and radius increases because:
- The valence electrons gets further from the nucleus and **reduces the force of attraction** of the nucleus **on an incoming electron**.
- The **core electron shells 'shield' the valence electron shell** from the attraction of the nucleus.

Halogens **react with Group 1 metals** to form **ionic halides** e.g. sodium chloride (NaCl), lithium bromide (LiBr).
Sodium loses an outer electron which is gained by chlorine.
- the metal gains a +1 charge.
- the halogen gains a -1 charge.

The oppositely charged ions are attracted to each other by **electrostatic attraction**.

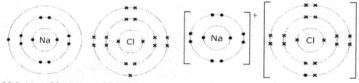

$2M\ (s) + X_2\ (g) \rightarrow 2MX\ (s)$
$2Na\ (s) + Cl_2\ (g) \rightarrow 2NaCl\ (s)$

A more reactive halogen is one which has a **stronger attraction for an electron** (more **electronegative**).
It will **displace** a less reactive halide from its compounds in solution i.e. it **oxidises the less reactive halide ion** to the halogen.
$2X^- + Y_2 \rightarrow 2Y^- + X_2$
X^-: less reactive halide
Y_2: more reactive halogen
Y^-: halogen is reduced – it is an oxidising agent
X_2: halide is oxidised

Example:

potassium bromide and chlorine - the colourless solution turns reddish-brown as bromine is liberated from its salt solution by the more reactive chlorine.
potassium bromide + chlorine → potassium chloride + bromine
2KBr (aq) + Cl$_2$ (aq) → 2KCl (aq) + Br$_2$ (aq)
N.B. Cl$_2$ (aq) is chlorine water (solution) and Br$_2$ (aq) is bromine water (solution).

Tests for halides
Add dilute HNO$_3$ followed by AgNO$_3$ solution - HNO$_3$ prevents the precipitation of other ions e.g. CO$_3^{2-}$.
General ionic equation: **Ag$^+$ (aq) + X$^-$ (aq) → AgX (s)**
The colour of the precipitate AgX identifies the halide:
o   AgCl – White
o   AgBr – Cream
o   AgI – Yellow

The solubility of the precipitate is tested with NH$_3$ solution (dilute then concentrated).
AgCl – precipitate **dissolves in dilute ammonia solution** to give a colourless solution.
AgBr – precipitate is almost unchanged using dilute ammonia solution, but **dissolves in concentrated ammonia solution** to give a colourless solution.
AgI – precipitate is **insoluble in ammonia** solution of any concentration.

Period 3 Oxides
Formulae
Going across period 3, there is a steady **increase in the number of oxygen atoms** that each element bonds to:
- the number of valence electrons available for bond formation increases.
- for the highest oxides, each successive element bonds to an extra half oxygen: Na_2O, MgO, Al_2O_3, SiO_2, P_4O_{10}, SO_3, Cl_2O_7

For the non-metals, **more than one oxide exists** and these contain the element in different oxidation states:
- phosphorus(III) oxide (P_4O_6), phosphorus(V) oxide (P_4O_{10}).
- sulfur dioxide (SO_2), sulfur trioxide (SO_3).
- dichlorine monoxide (Cl_2O), chlorine dioxide (ClO_2), dichlorine heptoxide (Cl_2O_7).

Structures
Groups 1-13: giant ionic – Na_2O (s), MgO (s) and Al_2O_3 (s)
- Na_2O and MgO – ionically bonded, solids with high melting and boiling points.
- Al_2O_3 – characteristics of both ionic and covalent bonding, exceptionally high melting point.

Group 14: giant covalent – SiO_2 (s)
- SiO_2 – giant covalent lattice (each atom is joined to all the others by strong covalent bonds), solid with very high melting and boiling points.

Groups 15-17: molecular covalent – P_4O_{10} (s), P_4O_6 (s), SO_3 (l), SO_2 (g), Cl_2O_7 (l) and Cl_2O (g)
- molecular covalent bonding.
- compounds are gases, liquids or low melting point solids as a result of the relatively weak forces that exist between the molecules.

Large difference in electronegativity between the elements leads to more **ionic** character – these are usually compounds formed between **metals and non-metals**.

Small difference in electronegativity between the elements leads to more **covalent** character – these are usually compounds formed **between non-metals**.

Acid-base character

Formulae (state under standard conditions)	Na₂O (s)	MgO (s)	Al₂O₃ (s)	SiO₂ (s)	P₄O₁₀ (s)	SO₃ (l) SO₂ (g)
Acid-base character	basic	basic	amphoteric	insoluble in water, weakly acidic	acidic	

Groups 1-2: Na₂O (s) and MgO (s) are **basic**. The oxide ion can form a bond to hydrogen ions and as a result these ionic oxides act as bases by:
- dissolving in water to give alkaline solutions.
  O²⁻ (s) + H₂O (l) → 2OH⁻ (aq)
  Na₂O (s) + H₂O (l) → 2NaOH (aq)
  MgO (s) + H₂O (l) → Mg(OH)₂ (aq)
- neutralising acids to produce a salt and water.
  O²⁻ (s) + 2H⁺ (aq) → H₂O (l)
  Na₂O (s) + 2HCl (aq) → 2NaCl (aq) + H₂O (l)
  MgO (s) + 2HCl (aq) → MgCl₂ (aq) + H₂O (l)
  Na₂O (s) + H₂SO₄ (aq) → Na₂SO₄ (aq) + H₂O (l)
  MgO (s) + H₂SO₄ (aq) → MgSO₄ (aq) + H₂O (l)

Group 13: Al₂O₃ (s) is **amphoteric** (reacts with and hence dissolves in, both acids and alkalis). It is insoluble in water.
- Aluminium oxide reacts with acids:
  Al₂O₃ (s) + 6H⁺ (aq) → 2Al³⁺ (aq) + 3H₂O (l)
  Al₂O₃ (s) + 3H₂SO₄ (aq) → Al₂(SO₄)₃ (aq) + 3H₂O (l)
  Al₂O₃ (s) + 6HCl (aq) → 2AlCl₃ (aq) + 3H₂O (l)
- Aluminium oxide reacts with bases:
  Al₂O₃ (s) + 2OH⁻ (aq) + 3H₂O (l) → 2Al(OH)₄⁻ (aq)
  Al₂O₃ (s) + 2NaOH (aq) + 3H₂O (l) → 2NaAl(OH)₄ (aq)

Moving towards the middle of the periodic table the ionization energy becomes too great for cation formation and the elements tend towards non–metallic behaviour. In this region the elements (P, Si and Cl) bond by means of covalent bonds.

- Carbon dioxide (CO_2) reacts reversibly with water to produce carbonic acid.

$$CO_2 \text{ (g)} + H_2O \text{ (l)} \rightleftharpoons H_2CO_3 \text{ (aq)}$$

- Sulfur trioxide (SO_3) reacts with water to produce sulfuric acid.

$$SO_3 \text{ (l)} + H_2O \text{ (l)} \rightarrow H_2SO_4 \text{ (aq)}$$

- Sulfur dioxide (SO_2) reacts reversibly with water to produce sulfurous acid.

$$SO_2 \text{ (s)} + H_2O \text{ (l)} \rightleftharpoons H_2SO_3 \text{ (aq)}$$

- Phosphorus oxide (P_4O_{10}) reacts with water to produce phosphoric acid.

$$P_4O_{10} \text{ (s)} + 6H_2O \text{ (l)} \rightarrow 4H_3PO_4 \text{ (aq)}$$

- Silicon dioxide (SiO_2) does not react with water (insoluble) but neutralises sodium hydroxide (NaOH) to form sodium silicate (Na_2SiO_3).

$$SiO_2 \text{ (s)} + 2NaOH \text{ (aq)} \rightarrow Na_2SiO_3 \text{ (aq)} + H_2O \text{ (l)}$$

13.1 First-row d-block elements
Syllabus
Nature of science – can you relate this topic to these concepts?
Looking for trends and discrepancies – transition elements follow certain patterns of behaviour. The elements Zn, Cr and Cu do not follow these patterns and are therefore
considered anomalous in the first-row d-block.
Understandings – how well can you explain these statements?
Transition elements have variable oxidation states, form complex ions with ligands, have coloured compounds, and display catalytic and magnetic properties.
Zn is not considered to be a transition element as it does not form ions with incomplete d-orbitals.
Transition elements show an oxidation state of +2 when the s-electrons are removed.
Applications and skills – how well can you do all of the following?
Explanation of the ability of transition metals to form variable oxidation states from successive ionization energies.
Explanation of the nature of the coordinate bond within a complex ion.
Deduction of the total charge given the formula of the ion and ligands present.
Explanation of the magnetic properties in transition metals in terms of unpaired electrons.

Transition metals
Found in the **d-block** of the periodic table between Groups 2 and 13.
Definition: **A transition element is an element that forms at least one ion with a partly filled d-orbital.**
Physical properties
- strong with high melting points due to metallic bonding.
- electrical and thermal conductivity due to delocalised electrons.

Chemical properties
- variable oxidation states e.g. Cu^+, Cu^{2+}.
- form coloured compounds.
- form complexes with ligands.
- often used as catalysts.

3	4	5	6	7	8	9	10	11	12
Sc	Ti	V	Cr	Mn	Fe	Co	Ni	Cu	Zn
Y	Zr	Nb	Mo	Tc	Ru	Rh	Pd	Ag	Cd
La	Hf	Ta	W	Re	Os	Ir	Pt	Au	Hg
Ac	Rh	Db	Sg	Gh	Hs	Mt	Ds	Rg	Cn

Group 12 elements are not considered transition elements: d-sublevels are full (contain 10 electrons)

Electronic configuration of atoms

Scandium (Sc)
[Ar] | ↑ | | | | | | ↑↓ |

Titanium (Ti)
[Ar] | ↑ | ↑ | | | | | ↑↓ |

Vanadium (V)
[Ar] | ↑ | ↑ | ↑ | | | | ↑↓ |

Chromium (Cr)
[Ar] | ↑ | ↑ | ↑ | ↑ | ↑ | | ↑ |

Manganese (Mn)
[Ar] | ↑ | ↑ | ↑ | ↑ | ↑ | | ↑↓ |

Iron (Fe)
[Ar] | ↑↓ | ↑ | ↑ | ↑ | ↑ | | ↑↓ |

Cobalt (Co)
[Ar] | ↑↓ | ↑↓ | ↑ | ↑ | ↑ | | ↑↓ |

Nickel (Ni)
[Ar] | ↑↓ | ↑↓ | ↑↓ | ↑ | ↑ | | ↑↓ |

Copper (Cu)
[Ar] | ↑↓ | ↑↓ | ↑↓ | ↑↓ | ↑↓ | | ↑ |

Zinc (Zn)
[Ar] | ↑↓ | ↑↓ | ↑↓ | ↑↓ | ↑↓ | | ↑↓ |

Scandium (Sc): [Ar] $3d^1 4s^2$
Titanium (Ti): [Ar] $3d^2 4s^2$
Vanadium (V): [Ar] $3d^3 4s^2$
Chromium* (Cr): [Ar] $3d^5 4s^1$
Manganese (Mn): [Ar] $3d^5 4s^2$
Iron (Fe): [Ar] $3d^6 4s^2$
Cobalt (Co): [Ar] $3d^7 4s^2$
Nickel (Ni): [Ar] $3d^8 4s^2$
Copper** (Cu): [Ar] $3d^{10} 4s^1$
Zinc (Zn): [Ar] $3d^{10} 4s^2$
* Extra stability of half-filled d-orbital.
** Extra stability of fully filled d-orbital.

Electronic configuration of ions

Vanadium (II), V^{2+}

[Ar] | ↑ | ↑ | ↑ | | | □

Chromium (I), Cr^+

[Ar] | ↑ | ↑ | ↑ | ↑ | ↑ | □

Manganese (II), Mn^{2+}

[Ar] | ↑ | ↑ | ↑ | ↑ | ↑ | □

Iron (II), Fe^{2+}

[Ar] | ↑↓ | ↑ | ↑ | ↑ | ↑ | □

Cobalt (I), Co^+

[Ar] | ↑↓ | ↑↓ | ↑ | ↑ | ↑ | ↑

Nickel (II), Ni^{2+}

[Ar] | ↑↓ | ↑↓ | ↑↓ | ↑ | ↑ | □

Copper (II), Cu^{2+}

[Ar] | ↑↓ | ↑↓ | ↑↓ | ↑↓ | ↑ | □

Zinc (II), Zn^{2+}

[Ar] | ↑↓ | ↑↓ | ↑↓ | ↑↓ | ↑↓ | □

Evidence for electronic congifurations
- **Slight jump** in the successive ionisation energies between removing the **last 4s-electron** and the **first 3d-electron**.
- **Big jump** in successive ionisation energy **after the last 3d-electron** has been removed. The next electron comes from a 3p-orbital and is more strongly attracted to the nucleus.

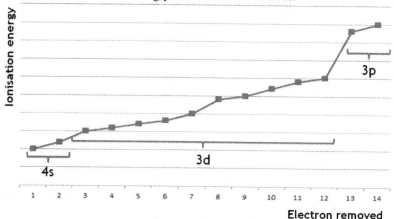

Oxidation states

Oxidation state is the number of electrons, that atoms have to lose or gain or share, when forming ionic or covalent bonds in compounds
- Oxidation state of an uncombined element is **zero.**
- For a monatomic ion the oxidation state equals the charge on the ion.
- With compounds of more than one element, the most electronegative element is given the negative oxidation state.
- Oxidation state of hydrogen is +1 except in metal hydrides.
- Oxidation state of oxygen is −2 except in peroxides, e.g. H_2O_2, when it is −1.
- The sum of all oxidation states in a neutral compound is zero.

Variable oxidation states
o Most common oxidation state is +2 – occurs when the two 4s electrons are lost.
o **Highest oxidation state** corresponds to the **loss of all 3d- and 4s- electrons.**
o 3d electrons are very close in energy to the 4s electrons
  o Many different ions formed.
  o Similar properties shown by transition elements.
o The **3d electrons** quite effectively **shield** the outer **4s electrons**, the first ionization energy remains relatively constant. As a result the d–block elements have many similar chemical and physical characteristics.

Most common oxidation states

Sc	Ti	V	Cr	Mn	Fe	Co	Ni	Cu	Zn
			+1	+1	+1	+1	+1	+1	
+2	+2	+2	+2	+2	+2	+2	+2	+2	+2
+3	+3	+3	+3	+3	+3	+3	+3	+3	
	+4	+4	+4	+4	+4	+4	+4		
		+5	+5	+5					
			+6	+6	+6				
				+7					

Stable high oxidation states ← +3 is the most common oxidation state
Stable low oxidation states → +2 is the most common oxidation state

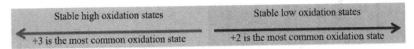 +x common oxidation state

Some specific examples:
o Chromium
  $CrCl_2$ (Cr = +2)
  $Cr_2(SO_4)$ (Cr = +3)
  $K_2Cr_2O_4$ and $K_2Cr_2O_7$ (Cr = +3 and +6)
o Iron
  $FeSO_4$ (Fe = +2)
  $FeCl_3$ (Fe = +3)
o Copper
  CuCl and Cu_2O (Cu = +1)
  $CuSO_4$ and CuO (Cu = +2)

The **3d and 4s** electrons have **similar energies**.

There is no sudden increase in successive ionization energies of the transition elements in the same way as there is with the s–block elements (until all the 3d and 4s electrons have been lost).

The **highest oxidation states** usually occur as **oxyanions**, such as dichromate(VI) ($Cr_2O_7^{2-}$) and **permanganate** (also referred to as manganate(VII) (MnO_4^-).

These tend to be **good oxidising agents**.

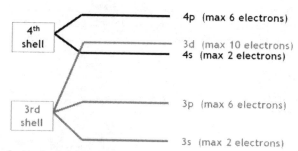

General physical properties
- high melting point.
- high electrical conductivity.
- high thermal conductivity.
- high tensile strength.
- malleable.
- ductile.

These properties are due to:
- **very strong metallic bonding** (all 4s and 3d electrons are involved as their energies are similar).
- a much larger number of **delocalised electron** per ion.

General chemical properties
- a variety of stable oxidation states in compounds.
- the ability to form complex ions.
- the formation of coloured ions.
- catalytic activity as elements or compounds.

Complexes and Ligands

Ligands are lone electron-pair donors, they are either **anions** or **neutral** molecules. The most common examples are water, ammonia, chloride ion and cyanide ion.

They form **dative** or **coordinate** covalent bonds with transition metal ions or atoms.

The ligand donates its electron pair to an **empty** 3d-, 4s- or 4p-orbital e.g.

$[Cu(H_2O)_6]^{2+}$

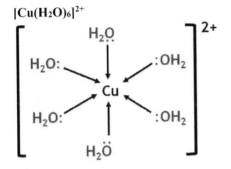

$[Ni(CN)_4]^{2-}$

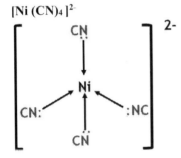

Coordination number

The number of **atoms, ions** or **groups** datively bonded with the central transition metal ion is known as the **coordination number.**

Complex	Coordination no.
$[Cu(H_2O)_6]^{2+}$	6
$[CoCl_4]^{2-}$	4
$[Ni(CN)_4]^{2-}$	4
$[Ag(NH_3)_2]^+$	2

Charge on the central ion

overall charge = charge on central ion + charge on ligands
therefore
charge on central ion = overall charge – charge on ligands

Transition metal complex	Charge on ligands	Overall charge	Charge on central ion
$[Fe(H_2O)_6]^{3+}$	6 × 0	+3	+3
$[CoCl_4]^{2-}$	4 × -1	-2	+2
$[Ni(CN)_4]^{2-}$	4 × -1	-2	+2
$[Ag(NH_3)_2]^+$	2 × 0	+1	+1

Shapes of Complexes

The stereochemistry (shape or structure) of transition metal complexes can be determined by X-ray crystallography which can show the bond angles and distances in the species.
Four-coordinate complexes are usually tetrahedral e.g. $[CoCl_4]^{2-}$.
Ions with d^8 configuration e.g. Pt^{2+}, Pd^{2+} and Au^{3+} are usually square planar e.g. $[Pt(NH_3)_2Cl_2]$.

Stereochemistry	Bond angles	Coordination number	Example
linear	180°	2	$K[Cr(Cl)_2]$
tetrahedral	109.5°	4	$K_2[CoCl_4]$
square planar	90° and 180°	4	$K_2[Ni(CN)_4]$, $[Pt(NH_3)_2Cl_2]$
octahedral	90° and 180°	6	$[Fe(H_2O)_6]Cl_3$

Common shapes shown by transition metal complexes:
M = transition metal ion
L = ligand
n+ = charge on ion

linear

$$\left[L{:}\!\longrightarrow\! M \!\longleftarrow\! {:}L \right]^{n+}$$

square planar

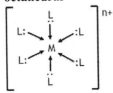

octahedral

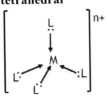

tetrahedral

$$\left[\begin{array}{c} L \\ \downarrow \\ L{:}\nearrow M \nwarrow {:}L \\ L{:} \end{array} \right]^{n+}$$

Four-coordinate complexes are usually tetrahedral e.g. $[CrCl_4]^{3-}$. Ions with d^8 configuration e.g. Pt^{2+}, Pd^{2+} and Au^{3+} are usually square planar e.g. $[Pt(NH_3)_2Cl_2]$.

Two-coordinate complexes are linear e.g. $[CrCl_2]^-$.

Types of ligands

Imagine that the ligands 'bite' onto the central ions. Ligands are classified using the Latin word *dentatus* meaning *having teeth*.
- **monodentate ligands**: form **one** coordinate bond with the central metal ion (bites once).

Examples: F^-, Cl^-, NH_3, H_2O, CN^-, OH^-
- **polydentate (chelate) ligands**: form **two or more** coordinate bonds with the central metal ion (bites many times).

Examples: 1,2-ethanediamine (en), ethanedioate $(ox)^{2-}$, ethylenediaminetetraacetate $(EDTA)^{4-}$

Uses of EDTA

Application	Chemical action
Heavy metal removal – Pb, Hg, As, etc. can be removed by administering calcium disodium EDTA, Na$_2$[Ca(EDTA)].	Heavy metal ions displace Ca^{2+} to form an anionic complex e.g. [Pb(EDTA)]$^{2-}$ which is passes by the kidneys into the urine. [Ca(EDTA)]$^{2-}$ + Pb^{2+} → [Pb(EDTA)]$^{2+}$ + Ca^{2+}
Chelation therapy – EDTA traps Ca^{2+}, preventing atherosclerotic tissue and plaque from forming.	Ca^{2+} + EDTA → [Ca(EDTA)]$^{2-}$ + 4H$^+$ Mg^{2+} + EDTA → [Mg(EDTA)]$^{2-}$ + 4H$^+$
Water softening, detergents and **shampoos** – EDTA traps Ca^{2+} and Mg^{2+} ions.	
Food preservation – Na$_2$[Ca(EDTA)] traps metal ions which catalyse radical reactions which cause rancidity in fats and oils.	**Hydrolytic rancidity**: lipids break down into fatty acids and glycerol (propan-1,2,3-triol). **Oxidative rancidity**: fatty acid chains are oxidised, oxygen added across C–C bonds in unsaturated lipids – volatile aldehydes and carboxylic acids are formed, which causes the 'rancid' odour.
Restorative sculpture – Brochantite, CuSO$_4$.3Cu(OH)$_2$ is an insoluble solid coating on brass and copper sculptures.	Brochantite reacts with EDTA to form [Cu(EDTA)]$^{2-}$ which is soluble and is washed away.
Preservative in **cosmetics**	Metal ions cause cosmetics to deteriorate in colour and fragrance properties.

Common complexes
- All transition metals form hexaqua hydrated ions e.g. [Cr(H$_2$O)$_6$]$^{3+}$, [Cu(H$_2$O)$_6$]$^{2+}$.
- Zinc forms [Zn(H$_2$O)$_4$]$^{2+}$.
- Iron (II) and iron (III) form complexes with six cyanide ions – [Fe(CN)$_6$]$^{4-}$ and [Fe(CN)$_6$]$^{3-}$.
- Copper (II) forms a complex with ammonia - [Cu(NH$_3$)$_4$(H$_2$O)$_2$]$^{2+}$.
- Chloride ligands are much larger than water molecules – the maximum number of chloride ligands around a metal ion is four e.g. [CuCl$_2$]$^-$ and [CuCl$_4$]$^{2-}$.

Catalytic activity of transition metals
Heterogeneous catalysts
Catalysts which are in a **different phase** from the reactants. Transition metals use their **d-orbitals** to provide **active sites** for the **adsorption** of reactants.

Examples
1. **Haber process** (production of ammonia): **iron** as catalyst. Ammonia is a raw material for fertilizers, plastics, drugs and explosives.

   $N_2 (g) + 3H_2 (g) \rightleftharpoons 2NH_3 (g)$

2. **Contact process** (manufacture of H_2SO_4 from SO_3): V_2O_5 **(vanadium (V) oxide)** as catalyst.

   $2SO_2 (g) + O_2 (g) \rightleftharpoons 2SO_3 (g)$

3. **Decomposition of hydrogen peroxide: manganese(IV) oxide (MnO_2)** as catalyst.

   $2H_2O_2 (aq) \rightarrow 2H_2O (l) + O_2 (g)$

4. **Hydrogenation of alkenes: nickel (Ni), palladium (Pd) or platinum (Pt)** as catalysts to convert (some) C=C double bonds in alkenes to C-C single bonds in the manufacture of margarine from vegetable oils.

$$\begin{array}{c}H\\H\end{array}\!\!C\!=\!C\!\!\begin{array}{c}H\\H\end{array} + H\text{-}H \rightarrow H\text{-}\underset{H}{\overset{H}{C}}\text{-}\underset{H}{\overset{H}{C}}\text{-}H$$

Hydrogenation of unsaturated oils: **nickel (Ni)** as catalyst turns liquid oils to semi-solid which increases usability (spreadable) and chemical stability.

RCH=CHR' (l) + H_2 (g) → RCH_2CH_2R'

Benefits:
- healthier for the heart than saturated fats e.g. lard and other animal fats.

Disadvantages:
- forms trans-fatty acids which:
  - are difficult to metabolise.
  - accumulate in the fatty tissues.
  - increase the levels of low-density lipoproteins (LDL) or 'bad' cholesterol – these narrow the arteries and lead to cardiovascular disease.

Advantages of heterogeneous catalysts
The catalyst can be **easily removed** by filtration from the reaction mixture. It can then be cleaned or processed to be **used again**.

Heterogeneous catalysts (catalytic converters)

- **Nitrogen monoxide (NO)** is produced when **nitrogen** gas from the air burns in the **intense heat (1500°C)** inside an **internal combustion engine**.

$N_2 (g) + O_2 (g) \rightarrow 2NO (g)$

- **Nitrogen dioxide (NO_2)** is produced when **NO** reacts with atmospheric **oxygen**.

$2NO (g) + O_2 (g) \rightarrow 2NO_2 (g)$

- NO_2 is **toxic** (causes **respiratory problems**), **photochemical smog** and **acid rain** by dissolving in rain water.

$3NO_2 (g) + H_2O (l) \rightarrow 2HNO_3 (aq) + NO (g)$

- CO is **toxic, odourless** and **colourless** which **impedes oxygen transport** by haemoglobin.
- Unburnt **ethane** and **propane** causes **ozone formation**.
- **Platinum (Pt), Palladium (Pd)** and **rhodium (Rh)** as catalysts convert NO, CO and hydrocarbons to N_2, CO_2 and H_2O.

$2NO (g) + 2CO (g) \rightarrow N_2 (g) + 2CO_2 (g)$
$CH_3CH_2CH_3 (g) + 5O_2 (g) \rightarrow 3CO_2 (g) + 4H_2O (g)$

Additional NO is produced due to the high temperatures which is **reduced to N_2 in a secondary chamber** by CuO or Cr_2O_3.

Homogeneous catalysts

Catalysts which are in the **same phase** as the reactants.
Transition metals have **variable oxidation states** and can catalyse many **redox reactions**. They work via an **intermediate compound or ion** (mixed effectively with the reactants) and are of particular importance in **biological processes**.

Examples
1. **Fe^{2+} in heme**: hemoglobin contains a heme group (central Fe^{2+} ion surrounded by four nitrogen atoms). Oxygen is transported through the bloodstream by bonding with the heme group.
   $O_2 + Fe^{2+}$ **(in heme group)** $\rightarrow O_2\text{-}Fe^{2+} \rightarrow O_2 + Fe^{2+}$
   O_2 attaches to Fe^{2+} as a monodentate ligand and forms an octahedral structure.
2. **Co^{3+} in vitamin B_{12}**: vitamin B_{12} contains an octahedral (six sites) Co^{3+} complex. Five sites are occupied by nitrogen atoms. The sixth site is responsible for the production of red blood cells and regulating the nervous system.

Paramagnetic properties
Transition metals which contain **unpaired electrons** can exhibit magnetic properties – they are **attracted to an external magnetic field**.

Diamagnetic properties
Transition metals which do not contain **unpaired electrons** can exhibit magnetic properties – they are **repelled by an external magnetic field**.

13.2 Coloured compounds
Syllabus
Nature of science – can you relate this topic to these concepts?
Models and theories – the colour of transition metal complexes can be explained through the use of models and theories based on how electrons are distributed in d-orbitals.
Transdisciplinary – colour linked to symmetry can be explored in the sciences, architecture, and the arts.
Understandings – how well can you explain these statements?
The d sub-level splits into two sets of orbitals of different energy in a complex ion.
Complexes of d-block elements are coloured, as light is absorbed when an electron is excited between the d-orbitals.
The colour absorbed is complementary to the colour observed.
Applications and skills – how well can you do all of the following?
Explanation of the effect of the identity of the metal ion, the oxidation number of the metal and the identity of the ligand on the colour of transition metal ion complexes.
Explanation of the effect of different ligands on the splitting of the d-orbitals in transition metal complexes and colour observed using the spectrochemical series.

Evolving theories on the formation of transition metal complexes

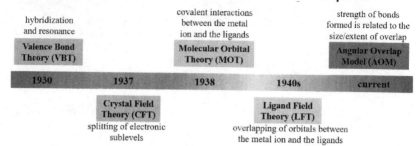

Characteristic colours shown by their compounds:
Blue: copper and cobalt.
Yellow: chromium.
Green: copper, nickel, chromium, iron.
Pink: manganese, cobalt.
Orange: iron, chromium.
Purple: titanium.

Reasons for coloured compounds

The colour of transition metal compounds are related to the presence of **partly filled** d orbitals.
Any ion with completely filled d orbitals is **not** coloured.
zinc compounds: Zn^{2+} $[Ar]3d^{10}4s^0$
Any ion with completely empty d orbitals is **not** coloured
titanium (IV) oxide: Ti^{4+} $[Ar]3d^04s^0$
Part of visible light is absorbed by a transition metal ion, part is either transmitted or reflected.
The wavelengths not absorbed give the colour of the compound.

Absorbed colour	λ/nm	Observed colour	λ/nm
Violet	400	Green-yellow	560
Blue	450	Yellow	600
Blue-green	490	Red	620
Yellow-green	570	Violet	410
Yellow	580	Dark Blue	430
Orange	600	Blue	450
Red	650	Green	520

The colour of light absorbed is complementary (diametrically opposed) to the colour of light transmitted.

Crystal Field Theory
When a ligand approaches the central metal ion, an **electrostatic field** is created by the ligand which alters the energy levels of the d-orbitals in two ways:
- **Isotropic** (spherically symmetrical) fields increase the energies uniformly – **degeneracy is preserved**.
- **Octahedral** fields **split the d orbitals** into two levels – t_{2g} (lower/stabilized) and e_g (higher/destabilized).

Octahedral field Splitting Pattern

The d-orbitals are split into an **upper group of two** and a **lower group of three**.

The value of the crystal field splitting energy, Δ_o (extent of splitting) depends on:
1. **Metal ion**
   - splitting **increases down a group** on the Periodic Table.
   Δ_o: $[Ni(H_2O)_6]^{2+} < [Pd(H_2O)_6]^{2+} < [Pt(H_2O)_6]^{2+}$
   - splitting **increases with higher oxidation states** – the ligands are attracted more closely to the central ion which increases the degree of overlap between the orbitals.
   Δ_o: $[V(H_2O)_6]^{2+} < [V(H_2O)_6]^{3+}$
2. **Ligand** – Spectrochemical series (section 15 in Data Booklet)
   - Ligands with a **high charge density** are known as **strong-field ligands**. They split the d-orbitals to a greater extent.
3. **Complex ion geometry**
   - The electric field created by the ligand's lone pair of electrons depends on the **geometry of the complex ion** e.g. tetrahedral complexes are split to a lesser extent than octahedral ones.

The **energy difference** (Δ_o) between the two levels corresponds to the **energy of a photon** in the **visible region** of the electromagnetic spectrum.

Spectrochemical series

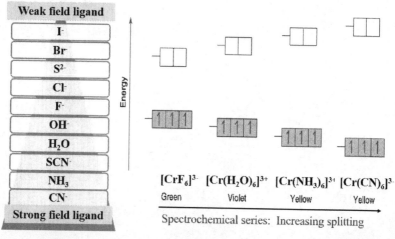

Crystal Field Theory (CFT) – Electronic configuration

The electrons in the split d orbitals are filled in the following order:

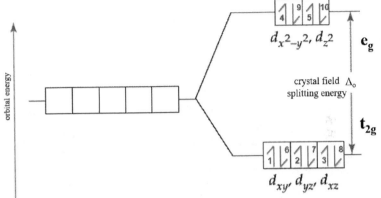

The fourth electron goes into the e_g level. For this electron to occupy the t_{2g} level, it must be provided with additional energy (**pairing energy, P**) to overcome repulsion by the three unpaired electrons. The electron configuration of the complex can be **spin-paired** or **spin-free** which depends on the extent of interaction with the attached ligands.

M^{2+} complexes

stronger-field ligands than NO_2^- will have **spin-paired** configurations e.g. $[Fe(CN)_6]^{4-}$

```
        ┌───┬───┐
       ─┤   │   ├─
        └───┴───┘
         $d_{x^2-y^2}, d_{z^2}$
        ┌───┬───┬───┐
        │ ↑↓│ ↑↓│ ↑↓│
        └───┴───┴───┘
         $d_{xy}, d_{yz}, d_{xz}$
```

weaker-field ligands than NO_2^- will have **spin-free** configurations e.g. $[Fe(H_2O)_6]^{2+}$

```
        ┌───┬───┐
       ─┤ ↑ │ ↑ ├─
        └───┴───┘
         $d_{x^2-y^2}, d_{z^2}$
        ┌───┬───┬───┐
        │ ↑↓│ ↑ │ ↑ │
        └───┴───┴───┘
         $d_{xy}, d_{yz}, d_{xz}$
```

M^{3+} complexes

stronger-field ligands than H_2O will have **spin-paired** configurations e.g. $[Cr(NH_3)_6]^{3+}$

```
        ┌───┬───┐
       ─┤   │   ├─
        └───┴───┘
         $d_{x^2-y^2}, d_{z^2}$
        ┌───┬───┬───┐
        │ ↑↓│ ↑ │ ↑ │
        └───┴───┴───┘
         $d_{xy}, d_{yz}, d_{xz}$
```

weaker-field ligands than H_2O will have **spin-free** configurations e.g. $[CrCl_6]^{3-}$

```
        ┌───┬───┐
       ─┤ ↑ │   ├─
        └───┴───┘
         $d_{x^2-y^2}, d_{z^2}$
        ┌───┬───┬───┐
        │ ↑ │ ↑ │ ↑ │
        └───┴───┴───┘
         $d_{xy}, d_{yz}, d_{xz}$
```

Crystal Field Theory (CFT) – Coloured complexes (d-to-d electronic transition)

When light shines through a complex ion, photons of a particular frequency are absorbed by an electron. The electron is promoted from the t_{2g} to the e_g level – this is called a **d-to-d electronic transition**.

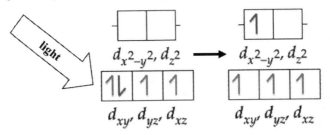

The energy of the photon transmitted, ΔE is related to Δ_o and its frequency or wavelength can be calculated from the following equation.

$$E = h\upsilon = \frac{hc}{\lambda}$$

h = Planck's constant = 6.63×10^{-34} Js
υ = frequency of radiation in Hz
c = speed of light = 3.00×10^8 ms^{-1}
λ = wavelength of radiation in m

4.1 Ionic bonding and structure
Syllabus
Nature of science – can you relate this topic to these concepts?
Use theories to explain natural phenomena – molten ionic compounds conduct electricity but solid ionic compounds do not. The solubility and melting points of ionic
compounds can be used to explain observations.
Understandings – how well can you explain these statements?
Positive ions (cations) form by metals losing valence electrons.
Negative ions (anions) form by non-metals gaining electrons.
The number of electrons lost or gained is determined by the electron configuration of the atom.
The ionic bond is due to electrostatic attraction between oppositely charged ions.
Under normal conditions, ionic compounds are usually solids with lattice structures.
Applications and skills – how well can you do all of the following?
Deduction of the formula and name of an ionic compound from its component ions, including polyatomic ions.
Explanation of the physical properties of ionic compounds (volatility, electrical conductivity and solubility) in terms of their structure.

Formation of ions

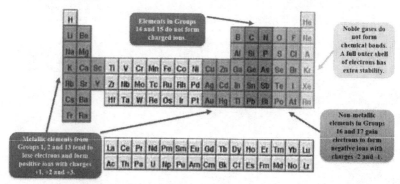

Atoms can have a tendency to gain or lose electrons to reach a noble gas electron configuration, which has added stability.

Oxidation is Loss (OIL)
Lithium loses an electron to achieve a [He] electron configuration.
Its oxidation state increases from 0 to +1 (it has been oxidised).
$$Li - e^- \rightarrow Li^+$$
Magnesium loses two electrons to achieve a [Ne] electron configuration.
Its oxidation state increases from 0 to +2 (it has been oxidised).
$$Mg - 2e^- \rightarrow Mg^{2+}$$

Reduction is Gain (RIG)
Bromine gains an electron to achieve a [Kr] electron configuration.
Its oxidation state decreases from 0 to -1 (it has been reduced).
$$Br + e^- \rightarrow Br^-$$
Oxygen gains two electrons to achieve a [Ne] electron configuration.
Its oxidation state decreases from 0 to -2 (it has been reduced).
$$O + 2e^- \rightarrow O^{2-}$$

Bond characteristics
Bond characteristics are dependent on the relative electronegativities (on the Pauling scale) of the elements. A difference in electronegativity of 1.8 units or more will result in a compound which is predominantly ionic.

H 2.1			increases				most electronegative
Li 1.0	Be 1.5	B 2.0	C 2.5	N 3.0	O 3.5	F 4.0	
Na 0.9	Mg 1.2	Al 1.5	Si 1.8	P 2.1	S 2.5	Cl 3.0	
K 0.8						Br 2.8	
Rb 0.8		least electronegative				I 2.5	

decreases ↓

Giant ionic lattice

The oppositely charged ions are attracted into a lattice that gets bigger and bigger until it consists of millions of ions. We have shown ions attracting and building into a 2 dimensional sheet. In fact the whole process will be going on in three dimensions to build up a giant 3-D lattice.

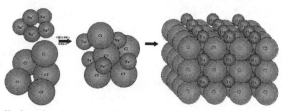

This is called a **giant ionic lattice**.

Cations and their charges

	1+ charge	2+ charge	3+ charge
Group 1	Li^+, Na^+, K^+, Rb^+, Cs^+		
Group 2		Be^{2+}, Mg^{2+}, Ca^{2+}, Sr^{2+}, Ba^{2+}	
Group 13			Al^{3+}
d-block	Ag^+ Cu^+ (in copper(I) compounds)	Zn^{2+} Mn^{2+} (in manganese(II) compounds) Fe^{2+} (in iron(II) compounds) Cu^{2+} (in copper(II) compounds)	Cr^{3+} (in chromium(III) compounds) Fe^{3+} (in iron(III) compounds)
Non-metal	NH_4^+ (ammonium)		

Anions and their charges

	1- charge	2- charge	3- charge
Group 17	F^- (fluoride) Cl^- (chloride) Br^- (bromide) I^- (iodide)		
Group 16		O^{2-} (oxide) S^{2-} (sulfide)	
Group 15			N^{3-} (nitride)
Polyatomic (oxoanions)	OH^- (hydroxide) NO_3^- (nitrate(V)) NO_2^- (nitrate(III)) HCO_3^- (hydrogencarbonate) MnO_4^- (manganate(VII)) OCl^- (chlorate(I)) ClO_3^- (chlorate(V)) CN^- (cyanide) O_2^- (superoxide)	O_2^{2-} (peroxide) CO_3^{2-} (carbonate) SO_4^{2-} (sulfate or sulfate(VI)) SO_3^{2-} (sulfite or sulfate(IV)) $C_2O_4^{2-}$ (ethanedioate) CrO_4^{2-} (chromate(VI)) $Cr_2O_7^{2-}$ (dichromate(VI))	PO_4^{3-} (phosphate(V)) PO_3^{3-} (phophonate)

Strength of ionic bonding (melting points of ionic compounds)
The oppositely charged ions are attracted by electrostatic forces to form an ionic bond. These forces need to be overcome to melt a solid ionic compound and this requires a very large input of energy. The strength of this bond or the attractive force is described by **Coulomb's law of electrostatics**:

$$|F| = \frac{Q_1 Q_2}{k(r_+ + r_-)^2}$$

$|F|$ = magnitude of force between the ions
$Q_1 Q_2$ = charges on the two ions
k = a constant
r_+, r_- = radii of the two ions

It can be seen from the equation that ionic bonds are **stronger** with:
- **bigger charges**
- **smaller ions**

Ionic compounds – other physical properties

Volatility
Ionic compounds have very low volatility due to the strength of the electrostatic forces between the oppositely charged ions.

Electrical conductivity
Conduct electricity only in the molten state – the ions must be freed from the solid lattice and be free to move.

Solubility
Most ionic compounds are soluble in polar solvents e.g. water. The partial charges of the solvent molecules ($\delta+$ and $\delta-$) are attracted to the charged ions. These solvent molecules pull the ions apart and surround them.

4.2 Covalent bonding
Syllabus
Nature of science – can you relate this topic to these concepts?
Looking for trends and discrepancies – compounds containing non-metals have different properties than compounds that contain non-metals and metals.
Use theories to explain natural phenomena – Lewis introduced a class of compounds which share electrons. Pauling used the idea of electronegativity to explain unequal sharing of electrons.
Understandings – how well can you explain these statements?
A covalent bond is formed by the electrostatic attraction between a shared pair of electrons and the positively charged nuclei.
Single, double and triple covalent bonds involve one, two and three shared pairs of electrons respectively.
Bond length decreases and bond strength increases as the number of shared electrons increases.
Bond polarity results from the difference in electronegativities of the bonded atoms.
Applications and skills – how well can you do all of the following?
Deduction of the polar nature of a covalent bond from electronegativity values.

Formation of covalent bonds
Formed by the:
- **sharing of electrons** (in pairs), each pair of shared electrons forms a bond (Lewis).
  1 pair → 1 bond
  2 pairs → 2 bonds
  3 pairs → 3 bonds
- **overlapping** of atomic **orbitals** → **molecular orbitals**.

Electronegativity

Electronegativity (χ_p): The ability of an element or atom to attract an electron pair in a covalent bond.
- **Decreases down a group.**
- **Increases across a period.**

If the difference in electronegativity between the elements is:
> **1.8** ; the bond is likely to be **mostly ionic**.
< **1.5** ; the bond is likely to be **mostly covalent**.

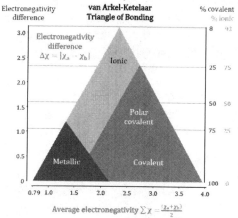

A pair of electrons that does not form a bond (non-bonding electrons) are called **lone-pair electrons**.

In a normal covalent bond, each atom donates one electron to the shared pair. In a **co-ordinate (dative)** bonds, electrons come from the same atom.

Lewis structures

When constructing Lewis structures, it is important that **all** the electrons in the shell are shown, including the non-bonding (lone) electrons. Electrons can be shown as dots or crosses.
A pair of lone electrons can be shown as two dots, two crosses or a single line.
A pair of bonding electrons can be shown as two dots, two crosses, a dot and a cross or a single line.

Hydrogen fluoride (HF)

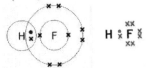

Water (H_2O)

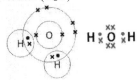

Methane (CH_4)

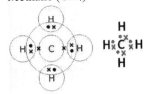

Ethene (C_2H_4)

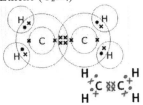

Nitrogen (N₂)

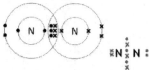

Oxygen (O₂)

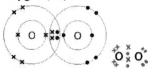

Ammonia (NH₃)

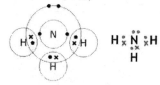

Beryllium chloride (BeCl$_2$)

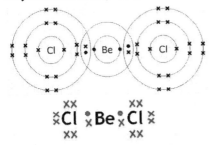

Boron trichloride (BCl$_3$)

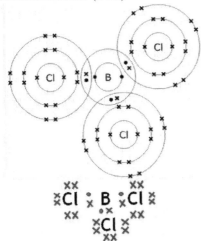

Beryllium and **boron** are metals but form **covalent bonds** because the elements have a **difference in electronegativity of less than 1.5**

Dative covalent (co-ordinate) bonding

- **Lewis theory** - both electrons in a covalent bond come from the same atom.
- **Molecular orbital theory** - overlap of an orbital with a pair of electrons with an empty orbital.

Dative bonds are shown with an **arrow**.

Hydronium ion, H_3O^+

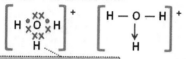

Both electrons are from O

Ammonium ion, NH_4^+

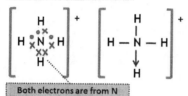

Both electrons are from N

Carbon monoxide, CO

:C::O:

Two bonds are formed by the sharing of electrons, one bond is dative.

Coordinate (dative covalent) bonding – Al₂Cl₆ dimer

Aluminium chloride (AlCl₃) melts under high pressure at 192.4°C and forms the Al₂Cl₆ **dimer** (two molecules joined together).
Aluminium is **tetravalent (coordination number 4)** with the central chlorine atoms on a perpendicular plane to the chlorine atoms at the edges.
The Al₂Cl₆ dimer is stable in the gaseous state up to 400°C and dissociates into AlCl₃ at higher temperatures.

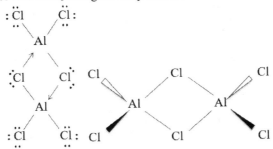

Covalent bond strength

The strength of covalent bonds is determined by:

o **Distance between the atoms** (sum of the atomic radii). Smaller atoms form stronger bonds.
Cl atom is smaller Br atom → C-Cl bond is shorter than C-Br bond → C-Cl bond is stronger

o **Number of electron pairs being shared**. Multiple bonds are stronger than single bonds.
Strength of bond: C≡C > C=C > C-C

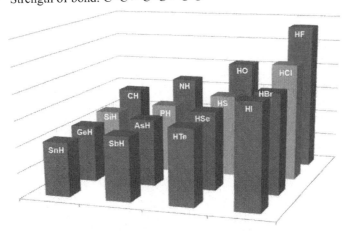

Bond polarity due to difference in electronegativity
When two covalently bonded atoms have different electronegativities, the shared pair of electrons are drawn towards the more electronegative atom.

This atom has a **partial negative charge (δ-)** the other tom has a **partial positive charge (δ+)**. The bond is **polar** or has a **dipole moment (μ)**.

δ+ δ-	δ+ δ-	δ+ δ-
H-----O	H-----Cl	C-----Cl

δ- δ+	δ- δ+
N-----H	C-----H

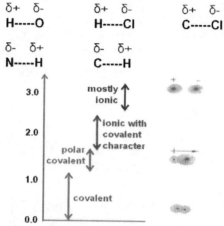

As an approximation, a **difference in electronegativity** of **more than 1.8** results in an **ionic bond**.

Dipole notation
Bond polarity can also be shown with an arrow and a plus sign.

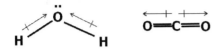

Microwaves in cooking
Microwave ovens emit waves with wavelengths ranging from 1 mm to 1 m (frequencies from 300 GHz to 300 MHz).
- The polar H_2O molecules absorb this radiation and flip their orientations with the same frequency.
- Energy is continually released due to this flipping in the form of heat and raises the temperature of the food.

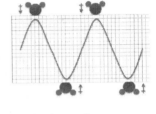

Polarity in molecules

A molecule which contains polar covalent bonds may not have a permanent dipole. The **overall polarity** depends on the **symmetry** of the molecule.

Symmetrical molecules – non-polar

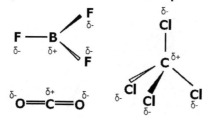

Asymmetrical molecules – polar

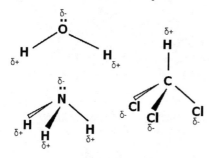

4.3 Covalent structures
Syllabus
Nature of science – can you relate this topic to these concepts?
Scientists use models as representations of the real world – the development of the model of molecular shape (VSEPR) to explain observable properties.

Understandings – how well can you explain these statements?
Lewis (electron dot) structures show all the valence electrons in a covalently bonded species.
The "octet rule" refers to the tendency of atoms to gain a valence shell with a total of 8 electrons.
Some atoms, like Be and B, might form stable compounds with incomplete octets of electrons.
Resonance structures occur when there is more than one possible position for a double bond in a molecule.
Shapes of species are determined by the repulsion of electron pairs according to VSEPR theory.
Carbon and silicon form giant covalent/network covalent structures.

Applications and skills – how well can you do all of the following?
Deduction of Lewis (electron dot) structure of molecules and ions showing all valence electrons for up to four electron pairs on each atom.
The use of VSEPR theory to predict the electron domain geometry and the molecular geometry for species with two, three and four electron domains.
Prediction of bond angles from molecular geometry and presence of non-bonding pairs of electrons.
Prediction of molecular polarity from bond polarity and molecular geometry.
Deduction of resonance structures, examples include but are not limited to C_6H_6, CO_3^{2-} and O_3.

Shapes of molecules
Sidgwick and Powell (1940):
1. Electron pair repulsions determined the shapes of molecules.
2. Electron pairs whether bonding or non-bonding repel each other and will arrange themselves in space to be as far apart as possible.

(Valence shell) Electron-pair repulsion theory (VSEPR)
How to work out the shape of the molecule:
1. Locate the central atom.
2. Order of Repulsion Strength: lone pair–lone pair > lone pair–double bond > lone pair–bonding pair > bonding pair–bonding pair
3. Geometry is determined by number of electron domains (charge centres) around the central atom. Single, double and triple bonds are counted as one electron domain. Each lone pair of electrons counts as one electron domain.

Molecules with 2 electron domains

Electron domain geometry: linear
Molecular geometry: linear
Bond angle: 180°

Beryllium chloride, $BeCl_2$

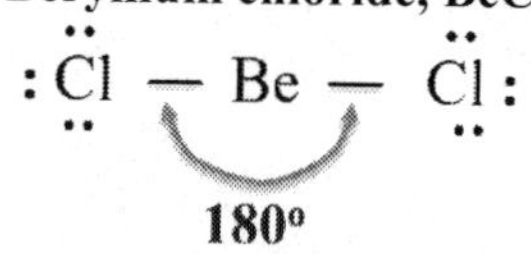

$:\ddot{C}l - Be - \ddot{C}l:$

180°

Carbon dioxide, CO_2

$:\ddot{O} = C = \ddot{O}:$

180°

Molecules with 3 electron domains
Electron domain geometry: trigonal planar
Molecular geometry: trigonal planar
Bond angle: 120°

Boron trifluoride, BF$_3$

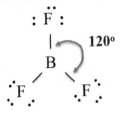

Nitrate ion, NO$_3^-$

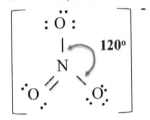

Electron domain geometry: trigonal planar
Molecular geometry: V-shaped (bent)
Bond angle: <120°

Sulfur dioxide, SO$_2$

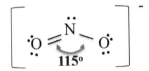

Nitrite ion, NO$_2^-$

Molecules with 4 electron domains
Electron domain geometry: tetrahedral
Molecular geometry: tetrahedral
Bond angle: 109.5°

Methane, CH_4

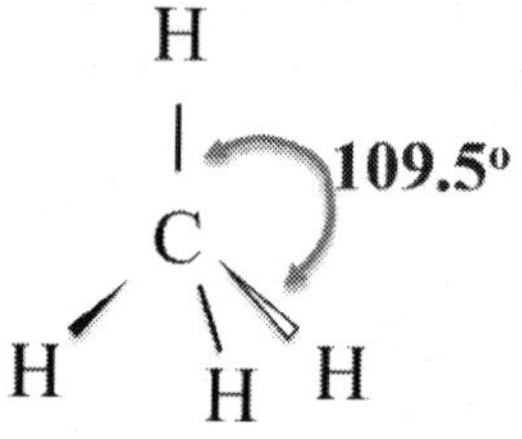

```
      H
      |  ⌒109.5°
      C
    /  \
  H    H  H
```

Ammonium ion, NH_4^+

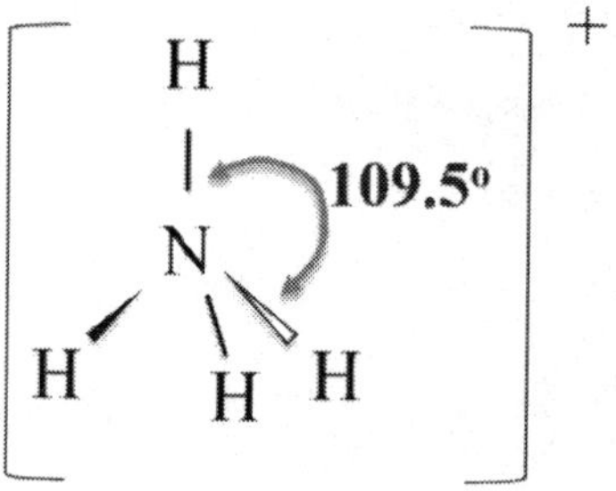

$$\left[\begin{array}{c} H \\ | \\ N \\ / | \backslash \\ H \quad H \quad H \end{array} \right]^+ \quad 109.5°$$

Electron domain geometry: tetrahedral
Molecular geometry: trigonal pyramidal
Bond angle: <109.5°

Ammonia, NH$_3$

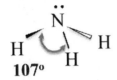

Sulfite ion, SO$_3^{2-}$

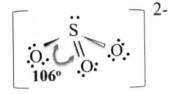

Hydronium ion, H$_3$O$^+$

Electron domain geometry: tetrahedral
Molecular geometry: V-shaped (bent)
Bond angle: <109.5°

Water, H₂O

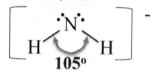

Amide ion, NH₂⁻

$$\left[\begin{array}{c} \ddot{\text{N}} \\ \text{H} \quad \text{H} \\ 105° \end{array} \right]^{-}$$

Delocalization of electrons and resonance structures
Electrons are said to be "delocalized" when they do not have a specific location:
- these electrons are **shared between more than one bonding position** in a molecule.
- they cannot be drawn in a simple Lewis structure.
- they exist in orbitals that include several atoms and/or bonds. You can imagine these orbitals as clouds surrounding parts of the molecule.
- they give molecules **resonance stability**.
- occurs in π–bonds where there is more than one possible location for a double bond.

Nitrate ion, NO_3^-
1. Calculate the number of valence electrons.
2. Draw the Lewis structure of the ion.

Three possible Lewis structures (**resonance forms**) exist:

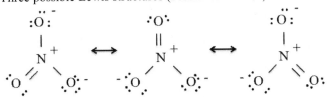

The structure of the species is a **resonance hybrid** (composite) of all three Lewis structures – the dotted lines are used to show **partial bonds**.

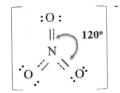

Nitrite ion, NO_2^-

Two possible Lewis structures (**resonance forms**) exist:

Sulfite ion, SO_3^{2-}
Three possible Lewis structures (**resonance forms**) exist:

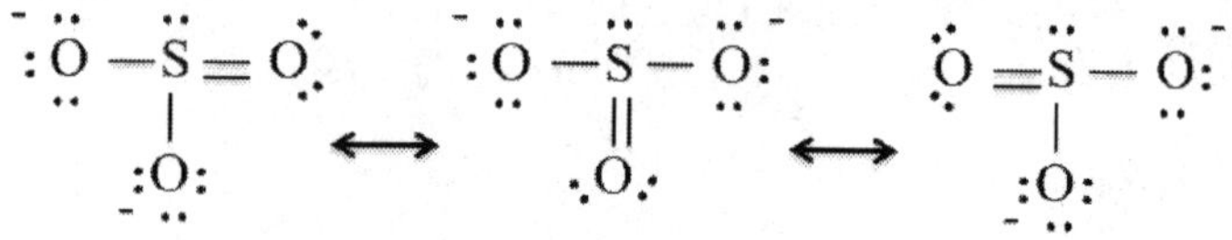

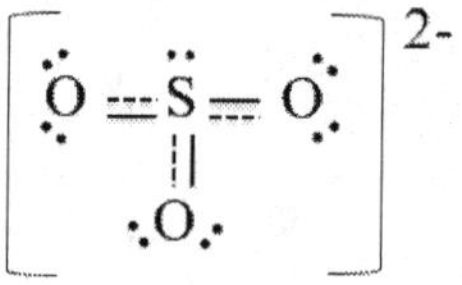

$$:\ddot{\underset{-}{O}}-\ddot{S}=\ddot{O}: \quad \longleftrightarrow \quad :\ddot{\underset{-}{O}}-\ddot{S}-\ddot{\underset{-}{O}}: \quad \longleftrightarrow \quad \ddot{O}=\ddot{S}-\ddot{\underset{-}{O}}:$$
$$\underset{:\ddot{\underset{-}{O}}:}{|} \qquad\qquad \underset{\cdot\ddot{O}\cdot}{\|} \qquad\qquad \underset{:\ddot{\underset{-}{O}}:}{|}$$

$$\left[\ddot{O}=\ddot{S}=\ddot{O} \atop \underset{\cdot\ddot{O}\cdot}{\|} \right]^{2-}$$

Carbonate ion, CO_3^{2-}

Three resonance structures shown, and the combined structure with dashed bonds:

$$\left[\begin{array}{c} :O: \\ \| \\ C \\ /\ \backslash \\ \ddot{O} \quad \ddot{O} \end{array} \right]^{2-}$$

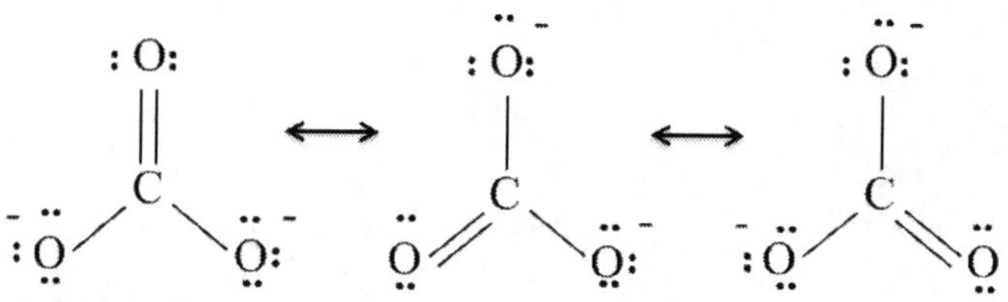

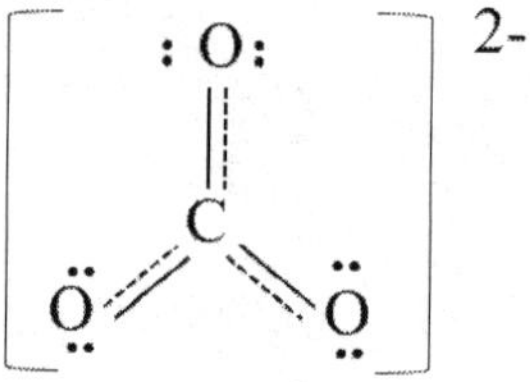

Ozone, O_3

Two resonance structures with O^+ center, and the combined structure with dashed bonds.

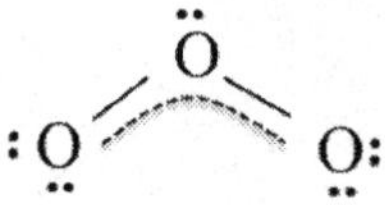

Methanoate ion, $HCOO^-$

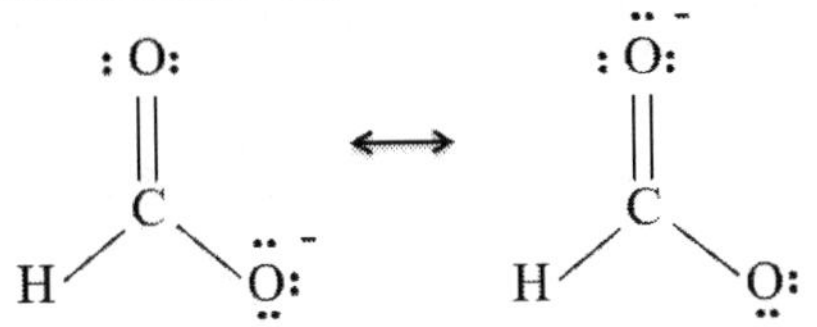

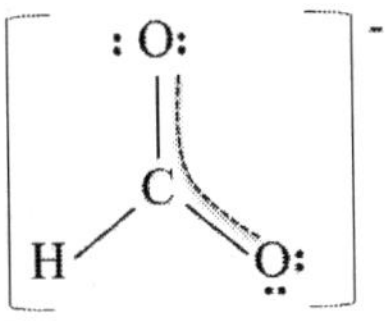

Ethanoate ion, CH_3COO^-

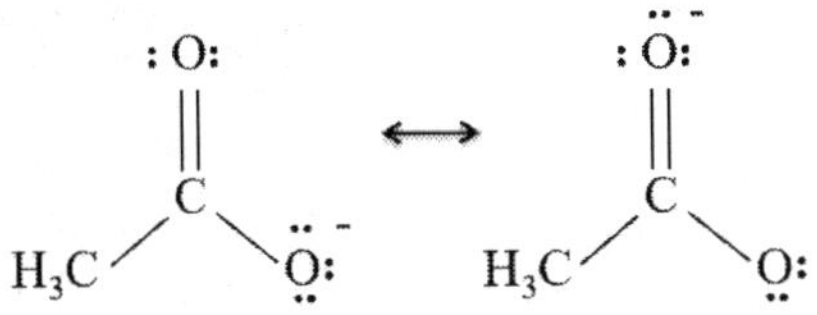

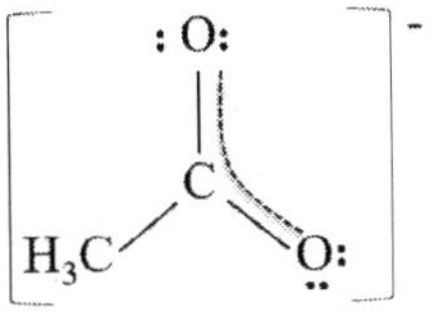

Benzene, C_6H_6

Michael Faraday isolated a new hydrocarbon in 1825. It was found to have the molecular formula C_6H_6 which suggests a large number of double bonds.

In 1865 after a dream about a snake biting its own tale, Kekulé suggested the following structure for benzene:

The structure of benzene was thought to be a resonance hybrid between two structures:

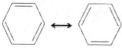

Benzene is now commonly represented as follows (the circle represents the delocalization of π-electrons):

Allotropes of Carbon – Covalent Network Solids (Diamond and Graphite)

In diamond, carbon atoms are arranged in a giant three-dimensional **tetrahedral** arrangement.

Each carbon atom is **covalently bonded** to **four** other carbon atoms with bond angles of **109.5°**.

In graphite, each carbon atom is **covalently bonded** to **three** other carbon atoms in a **layered hexagonal ring** structure with bond angles of **120°.**

The layers are **weakly connected** by **London forces** and can **slide over each other** – graphite breaks easily and is used in **pencils**.

The fourth π-electron is **delocalised** – graphite **conducts electricity**.

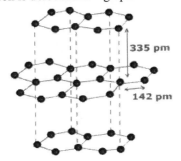

Both allotropes have **very high melting points** due to the **strong covalent bonds** between the atoms.

Allotropes of Carbon – Covalent Network Solids (Graphene)
Planar sheet of carbon atom hexagons. Each carbon atom is **covalently bonded** to **three** other carbon atoms in a **honeycomb lattice** structure with bond angles of **120°** and only one atom thick.

Uses of graphene:
- **Aerospace industry**: metal substitute due to its **low density** and **high strength** (100 times stronger than steel).
- **Liquid crystal displays (LCD)**: can be used to make **flexible touch displays** due to its **flexibility, transparency** and **electrical conductivity** (300 times better than copper).
- **Bioimaging**: graphene nanoribbons (GNRs) act as **contrast agents** for ultra high definition bioimaging.

Allotropes of Carbon (Fullerenes)
Fullerenes are formed when vaporized carbon condensed into clusters of 60 or 70 atoms. Each carbon atom is **covalently bonded** to **three** other carbon atoms and molecules have weak **London forces** between one another. The fourth electron is **delocalised** within the fullerene molecule but cannot travel to the next molecule. The C_{60} molecule (buckminsterfullerene) has the same basic geometry as a **geodesic dome** or a **truncated icosahedral cage** – 20 interlocking **hexagons** and 12 **pentagons**. Fullerenes are **black solids** and are **insoluble in water** but **soluble in some non-polar solvents** to form **coloured solutions**.

Uses of fullerenes:
- **Inclusion complexes**: genes and drugs can be carried by the fullerene molecules.
- **Radiation therapy**: radioactive radon-224 atoms can be trapped inside C_{60} molecules and emit α-rays which destroy cancer cells.
- **Ferromagnetism** and **superconductivity**: inclusion complexes can have unusual electromagnetic properties.
- **HIV treatment**: fullerenes can fit into the hydrophobic cavity in the active site of the HIV protease enzymes and inhibit its activity.

Allotropes of Carbon (Nanotubes)

Cylinder of interlocking hexagons of carbon atoms with diameters in the order of 10^{-9} mm.

One end of the tube can be sealed with a fullerene molecule.

Nanotubes have extraordinary **electrical conductivity** (10 times better than copper) and **strength** (100 times stronger than steel).

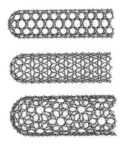

Uses of nanotubes:
- **Drug delivery**: nanotubes can be used to deliver drugs to specific parts of the body.
- **Electronics**: replaces silicon in many applications.
- **Synthesis of unstable substances**: these substances are stabilised within the environment of the nanotubes.

Covalent Network solid – silicon dioxide/silica (quartz)

Each Si atom is covalently bonded to four O atoms. Each O atom is covalently bonded to two Si atoms. This forms a dense **lattice of SiO$_4$ tetrahedra**.

SiO_2 has extremely high melting (1710°C) and boiling (2230°C) points due to the covalent bonds.

SiO_2 is **insoluble in water** and **does not conduct** electricity or heat in its solid crystalline state.

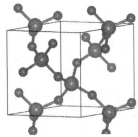

4.4 Intermolecular forces
Syllabus
Nature of science – can you relate this topic to these concepts?
Obtain evidence for scientific theories by making and testing predictions based on them – London (dispersion) forces and hydrogen bonding can be used to explain special interactions. For example, molecular covalent compounds can exist in the liquid and solid states. To explain this, there must be attractive forces between their particles which are significantly greater than those that could be attributed to gravity.

Understandings – how well can you explain these statements?
Intermolecular forces include London (dispersion) forces, dipole-dipole forces and hydrogen bonding.
The relative strengths of these interactions are:
London (dispersion) forces < dipole-dipole forces < hydrogen bonds.

Applications and skills – how well can you do all of the following?
Deduction of the types of intermolecular force present in substances, based on their structure and chemical formula.
Explanation of the physical properties of covalent compounds (volatility, electrical conductivity and solubility) in terms of their structure and intermolecular forces.

Summary of interactions between atoms, molecules and ions

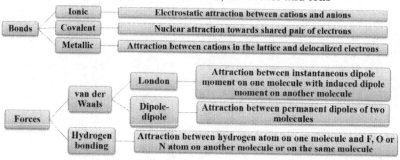

London (dispersion) forces
Exists between **all molecules**.
Electrons are constantly moving and at very high speeds.
At any one time, it is possible the one side has more electrons (the electron cloud is denser here).

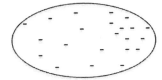

The imbalance of electrons density leads to the molecule having **partial charges** at either end – it has acquired an instantaneous dipole.

This **temporary dipole** then induces the same imbalance in its neighbouring molecules – they are attracted to one another.
The extent of interaction between the molecules depends on the ease of distortion of the electron clouds (**polarizability**).

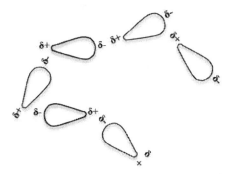

London (dispersion) forces - Factors affecting
1. Number of electrons
Atoms or molecules with more electrons are more polarizable as the electrons in the higher energy levels are further away from the nucleus and are less attracted to it. The potential energy, V associated with London forces is related to the distance, r by:

$$V \propto \frac{1}{r^6}$$

This can be seen in the increase in boiling points of Group 17 as we descend the group.

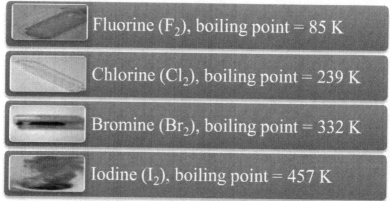

Fluorine (F_2), boiling point = 85 K

Chlorine (Cl_2), boiling point = 239 K

Bromine (Br_2), boiling point = 332 K

Iodine (I_2), boiling point = 457 K

2. Size of electron cloud
Straight chain alkanes
The London forces of attraction increase with the size of molecules. More electrons mean that the molecules are more polarisable as the electrons are further away from the nuclei and more contact points between the chains.

The pattern here is less regular as the smaller alkanes pack themselves into a crystal structure in the solid state.

Alkane	Methane (CH_4)	Ethane (C_2H_6)	Propane (C_3H_8)	Butane (C_4H_{10})	Pentane (C_5H_{12})	Hexane (C_6H_{14})
Melting point/K	91.1	89.8	83.4	135	143	178
Boiling point/K	109	185	231	273	309	342

3. Shapes of molecules
Branched alkanes
Branched chain alkanes usually have lower melting and boiling points than their straight chain isomers. The forces of attraction decrease because there are less contact points between the molecules.

2,2-dimethylpropane has a higher melting point. It has a symmetrical structure and can pack well in a solid crystal lattice.

Alkane	Pentane (C_5H_{12})	2-methylbutane	2,2-dimethylpropane
Melting point/K	143	113	257
Boiling point/K	309	301	283

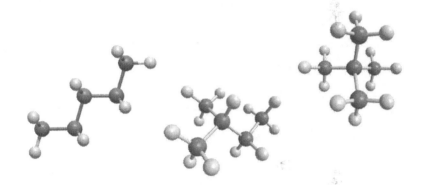

Dipole-dipole attraction
Permanent dipoles exist in polar molecules due to the **asymmetrical distribution of electrons** throughout the molecule e.g. HCl

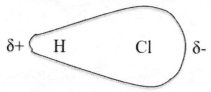

Symmetrical molecules cancel out any charge which will have risen from having different atoms joined along covalent bonds (**no net polarity**)
e.g. CH_4

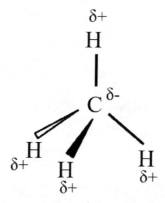

The dipoles line up so that the δ+ end of one molecule is next to the δ- end of another molecule.

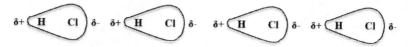

Hydrogen bonding

Exist between two partial electric charges of opposite polarity. **H** provides the positive charge ($\delta+$). This attracts a **lone pair of electrons** on another small bonded atom, usually **F, O,** or **N**. Cl atoms are too large.

Hydrogen bonding is essentially a type of permanent dipole-dipole force.

Type of bond	Bond energy/kJ mol^{-1}
Covalent	100 to 900
Hydrogen	20 to 50
Dipole-dipole	< 20
London	< 10

Hydrogen bonding through fluorine

Hydrogen fluoride (HF)

Hydrogen bonding through oxygen
Water (H$_2$O)

$$\begin{array}{c}
\cdots \overset{\delta-}{O} \cdots \\
\overset{\delta+}{H} \diagup \quad \diagdown \overset{\delta+}{H} \cdots \\
\qquad \qquad \cdots \overset{\delta-}{O} \cdots \\
\qquad \overset{\delta+}{H} \diagup \quad \diagdown \overset{\delta+}{H} \cdots
\end{array}$$

Hydrogen bonding explains why the **boiling point of water is higher** than the other Group 16 hydrides.

The density of ice is less than that of liquid water.
- Ice crystals contain interlocking rings of six H$_2$O molecules held together by hydrogen bonds.
- The distance between the molecules is larger in the solid than in the liquid.

Ethanol (C_2H_5OH) (liquid)

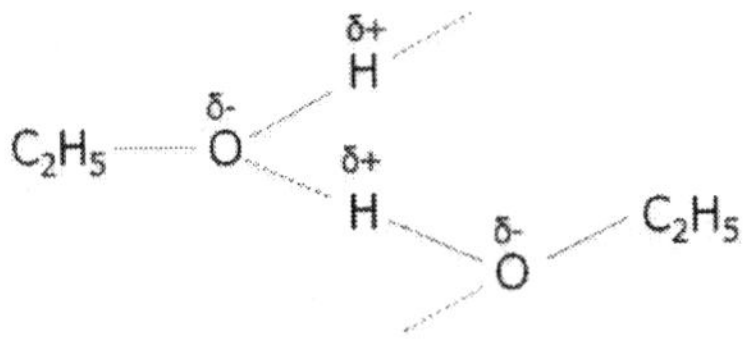

$$C_2H_5 \overset{\delta-}{-} O \cdots \overset{\delta+}{H} \quad \overset{\delta+}{H} \cdots \overset{\delta-}{O} - C_2H_5$$

Ethanol (C_2H_5OH) in water

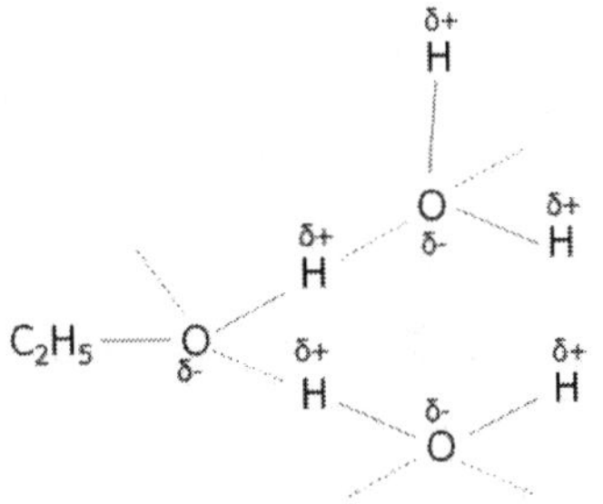

Ethanoic acid (CH_3COOH)

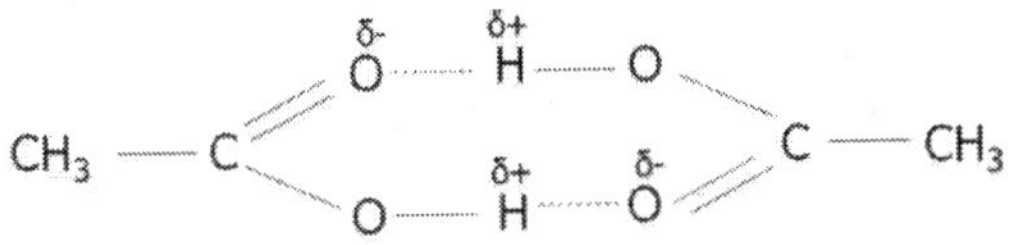

$$CH_3 - C \underset{O \cdots H \cdots O}{\overset{O \cdots H \cdots O}{\rightleftharpoons}} C - CH_3$$

This structure is called a **dimer**.

Hydrogen bonding through nitrogen

Ammonia (NH_3)

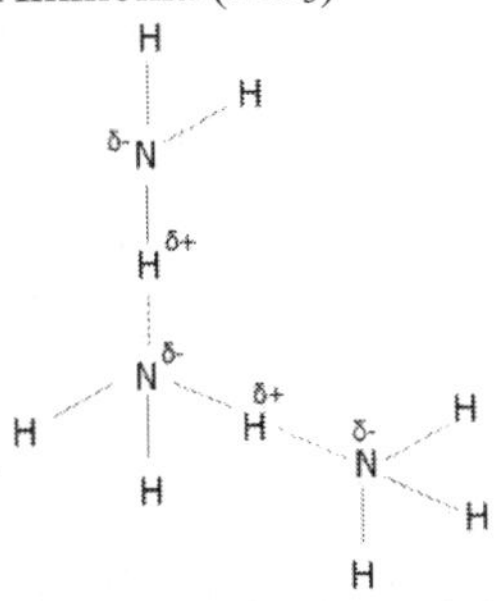

Other compounds which form hydrogen bonding include amines (with $-NH_2$ and $=N-H$ groups) and amino acids.

Proteins e.g. DNA molecules are long chains of amino acids held together by hydrogen bonds.

Physical properties due to bonding and intermolecular forces – Group 14 hydrides

- Intramolecular bonds are **covalent**.
- Intermolecular forces are **London forces**. All the hydrides are non-polar molecules because they are symmetrical.
- None of the elements in group 14 is electronegative enough to give hydrogen bonding.

Hydride	Methane (CH_4)	Silane (SiH_4)	Germane (GeH_4)	Stannane (SnH_4)
Boiling point/K	109	161	183	221

The **forces of attraction increase** with:
- **Size of molecules** – more electrons mean that the molecules are more polarisable.

Physical properties due to bonding and intermolecular forces – Group 15, 16 and 17 hydrides
- Intramolecular bonds are **covalent**.
- Intermolecular forces are **London forces** and **permanent dipole-dipole**. All the hydrides are **polar** molecules due to the presence of **lone pair electrons** on the more electronegative atom.
- The first hydride of each group (NH_3, H_2O and HF) form **hydrogen bonding** which increases their boiling points unexpectedly.

Group 5 hydride	NH_3	PH_3	AsH_3	SbH_3
Boiling point/K	**240**	185	218	256

Group 6 hydride	H_2O	H_2S	H_2Se	H_2Te
Boiling point/K	**373**	212	232	269

Group 7 hydride	HF	HCl	HBr	HI
Boiling point/K	**293**	188	206	238

Alcohols
Alcohols form hydrogen bonds between the molecules.
They are **less volatile than alkanes** of similar sizes and have **higher boiling points**.

Alkane	Ethane (CH_3CH_3)	Propane ($CH_3CH_2CH_3$)	Butane ($CH_3CH_2CH_2CH_3$)
M_r	30	44	58
Boiling point/K	185	231	273

Alcohol	Methanol (CH_3OH)	Ethanol (CH_3CH_2OH)	Propan-1-ol ($CH_3CH_2CH_2OH$)
M_r	32	46	60
Boiling point/K	338	352	371

4.5 Metallic bonding
Syllabus
Nature of science – can you relate this topic to these concepts?
Use theories to explain natural phenomena – the properties of metals are different from covalent and ionic substances and this is due to the formation of non-directional bonds with a "sea" of delocalized electrons.

Understandings – how well can you explain these statements?
A metallic bond is the electrostatic attraction between a lattice of positive ions and delocalized electrons.

The strength of a metallic bond depends on the charge of the ions and the radius of the metal ion.

Alloys usually contain more than one metal and have enhanced properties.

Applications and skills – how well can you do all of the following?
Explanation of electrical conductivity and malleability in metals.
Explanation of trends in melting points of metals.
Explanation of the properties of alloys in terms of non-directional bonding.

Drude-Lorentz model
Metal atoms **lose their valence electron(s)** and form **cations**.
- The cations are arranged in a **regular lattice** and surrounded by **delocalized electrons**.
- The positive ions are attracted to the sea of mobile electrons. These electrostatic attractions bind the entire crystal together as a single unit.

These electrons are delocalised around the cation lattice

The strength of metallic bonding increases with:
- Charge on metal ion.
- Number of valence electrons that can be delocalized.
- Smaller ionic radius of the metallic positive ion (cation).

Physical properties
Malleability and Ductility
Malleable: the ability to be pounded or hammered into different shapes without breaking.
Ductile: the ability to be stretched into long shapes or wires.

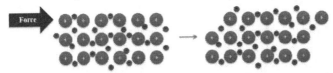

The metal cations slide over each other into new lattice positions (the attraction to the delocalized electrons maintain the lattice structure). Metallic bonding is said to be **non-directional** (acting in all directions) so that it stays intact even when the shape of the object changes.

Alloys

An alloy is a mixture of metals or metals and non-metals. Its physical properties can be changed in a number of ways:
- increased strength and durability e.g. iron + carbon + tungsten → steel
- increased resistance to corrosion e.g. iron + carbon + chromium → stainless steel
- increased magnetic properties e.g. iron + nickel + cobalt
- increased ductility e.g. copper + chromium + zirconium
- appearance e.g. copper + zinc → brass (which is a colour between the two constituent metals)

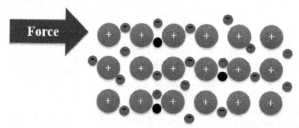

The carbon atom disrupt the layered structure of the metallic lattice and prevents the layer of iron atoms from sliding across each other, increasing its resistance to an external force.

Electrical conductivity

The delocalized electrons move through the metallic structure and can carry a current. The conductivity is increased when the metal is very pure (fewer impurities to disrupt the movement of electrons).

Trends in melting points

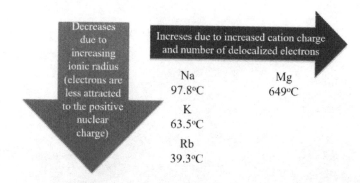

14.1 Further aspects of covalent bonding and structure
Syllabus
Nature of science – can you relate this topic to these concepts?
Principle of Occam's razor – bonding theories have been modified over time. Newer theories need to remain as simple as possible while maximizing explanatory power, for example the idea of formal charge.
Understandings – how well can you explain these statements?
Covalent bonds result from the overlap of atomic orbitals. A sigma bond (σ) is formed by the direct head-on/end-to-end overlap of atomic orbitals, resulting in electron density concentrated between the nuclei of the bonding atoms. A pi bond (π) is formed by the sideways overlap of atomic orbitals, resulting in electron density above and below the plane of the nuclei of the bonding atoms.
Formal charge (FC) can be used to decide which Lewis (electron dot) structure is preferred from several. The FC is the charge an atom would have if all atoms in the molecule had the same electronegativity. FC = (Number of valence electrons)-½(Number of bonding electrons)-(Number of non-bonding electrons). The Lewis (electron dot) structure with the atoms having FC values closest to zero is preferred.
Exceptions to the octet rule include some species having incomplete octets and expanded octets.
Delocalization involves electrons that are shared by/between all atoms in a molecule or ion as opposed to being localized between a pair of atoms.
Resonance involves using two or more Lewis (electron dot) structures to represent a particular molecule or ion. A resonance structure is one of two or more alternative Lewis (electron dot) structures for a molecule or ion that cannot be described fully with one Lewis (electron dot) structure alone.
Applications and skills – how well can you do all of the following?
Prediction whether sigma (σ) or pi (π) bonds are formed from the linear combination of atomic orbitals.
Deduction of the Lewis (electron dot) structures of molecules and ions showing all valence electrons for up to six electron pairs on each atom.
Application of FC to ascertain which Lewis (electron dot) structure is preferred from different Lewis (electron dot) structures.
Deduction using VSEPR theory of the electron domain geometry and molecular geometry with five and six electron domains and associated bond angles.
Explanation of the wavelength of light required to dissociate oxygen and ozone.
Description of the mechanism of the catalysis of ozone depletion when catalysed by CFCs and NO_x.

Formal Charge

The **formal charge** is the charge that the atom will have if all the atoms in the molecule had the same electronegativity.

Formal Charge (FC) = (no. of valence electrons) – ½ (number of bonding electrons) – (number of non-bonding electrons)

$$FC = V - \frac{B}{2} - N$$

Carbon tetrafluoride, CF_4
FC (C) = 4 – (8/2) – 0 = 0
FC (F) = 7 – (2/2) – 6 = 0
Net charge = 0 + 4(0) = 0

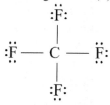

Ozone, O_3
FC (O_A) = 6 – (4/2) – 4 = 0
FC (O_B) = 6 – (6/2) – 2 = +1
FC (O_C) = 6 – (2/2) – 6 = -1
Net charge = 0 + (+1) + (-1) = 0

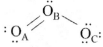

Nitrate ion, NO_3^-
FC (N) = 5 – (8/2) – 0 = +1
FC (O_A) = 6 – (2/2) – 6 = -1
FC (O_B) = 6 – (4/2) – 4 = 0
Net charge = +1 + 2(-1) + 0 = -1

Carbonate ion, CO_3^{2-}
FC (C) = 4 − (8/2) − 0 = 0
FC (O_A) = 6 − (2/2) − 6 = −1
FC (O_B) = 6 − (4/2) − 4 = 0
Net charge = +2 + 2(−2) + 0 = −2

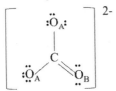

Formal Charge − incomplete octets (fewer than 8 electrons)
In cases where more than one Lewis structure is possible, the one which is chosen will have:
- Difference in FC, $\Delta FC = FC_{max} - FC_{min}$ closest to 0.
- Negative charges on the most electronegative atoms.

Boron trifluoride, BF_3
Structure 1
FC (B) = 3 − (6/2) − 0 = 0
FC (F) = 7 − (2/2) − 6 = 0
ΔFC = 0 − 0 = 0

:F:
|
B
/ \
:F: :F:

Structure 2
FC (B) = 3 − (8/2) − 0 = −1
FC (F_A) = 7 − (2/2) − 6 = 0
FC (F_B) = 7 − (4/2) − 4 = +1
ΔFC = +1 − (−1) = +2

:F_A:
|
B
// \
:F_B :F_A:

Structure 1 is the correct structure even though boron has an incomplete octet (6 electrons).

Formal Charge – expanded octets (more than 8 electrons)

If the central atom is an element from the 3rd period or below, the compound can have more than 8 electrons (4 pairs) around the central atom.

Phosphorus pentachloride, PCl$_5$

Phosphorus in covalent compounds can expands its octet, enabling ten valance electrons in 5 electron domains.

Electron domain geometry: trigonal bipyramidal
Molecular geometry: trigonal bipyramidal
Bond angles: 120° in the plane, 90° between the equatorial and axial atoms

If the central atom has non-bonding pairs of electrons, they occupy the equatorial positions in the molecule.

Sulfur tetrafluoride, SF$_4$

Sulfur expands its octet, enabling ten valance electrons in 5 electron domains.

Electron domain geometry: trigonal bipyramidal
Molecular geometry: see-saw
Bond angles: <120° between the equatorial S-F bonds, <90° between the equatorial plane and the axial S-F bonds

If the central atom has non-bonding pairs of electrons, they occupy the equatorial positions in the molecule.

Chlorine Trifluoride, ClF$_3$

Chlorine expands its octet, enabling ten valance electrons in 5 electron domains.

$$:\!\ddot{\underset{..}{F}}\!-\!\overset{<90°}{\underset{|}{\underset{:\!\ddot{\underset{..}{F}}\!:}{Cl}}}\!\overset{:\ddot{\underset{..}{F}}:}{}\!\ddot{:}$$

Electron domain geometry: trigonal bipyramidal
Molecular geometry: T-shaped
Bond angles: <120° between the equatorial F-Cl bonds, <90° between the equatorial plane and the axial F-Cl bonds

If the central atom has non-bonding pairs of electrons, they occupy the equatorial positions in the molecule.

Triiodide ion, I$_3^-$

Iodine expands its octet, enabling ten valance electrons in 5 electron domains.

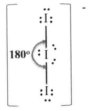

Electron domain geometry: trigonal bipyramidal
Molecular geometry: linear
Bond angle: 180°

Sulfur hexafluoride, SF_6
Sulfur expands its octet, enabling twelve valance electrons in 6 electron domains.

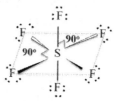

Electron domain geometry: octahedral
Molecular geometry: octahedral
Bond angles: 90°

In some cases, the non-bonding pairs of electrons can occupy the axial positions in the molecule.

Bromine pentafluoride, BrF_5
Bromine expands its octet, enabling twelve valance electrons in 6 electron domains.

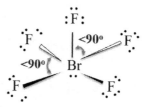

Electron domain geometry: octahedral
Molecular geometry: sqaure-based pyramidal
Bond angles: <90°

Xenon tetrafluoride, XeF_4
Xenon expands its octet, enabling twelve valance electrons in 6 electron domains.

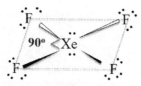

Electron domain geometry: octahedral
Molecular geometry: square planar
Bond angles: 90°

Molecular orbital theory (MOT) – Bonding orbitals

A covalent bond is formed when two atomic orbitals, each containing a single unpaired electron overlaps to form a molecular orbital which is at a **lower energy**. An increase in energy leads to **antibonding molecular orbitals**.

Two types of bonding molecular orbitals are possible: **σ-bond** and **π-bond**.

Sigma (σ)-bonding
The orbitals **overlap head on**.

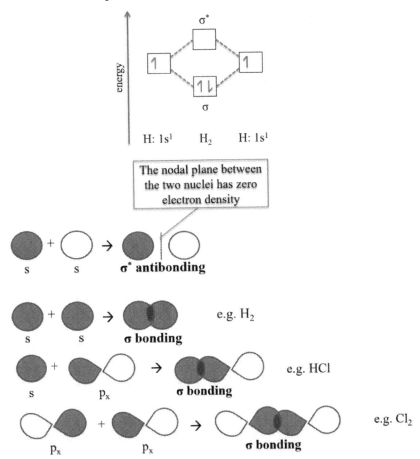

Pi (π)-bonding
The orbitals and **overlap sideways**.

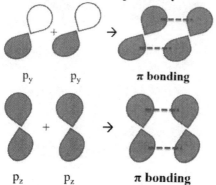

Molecular orbital theory (MOT) – Antibonding orbitals

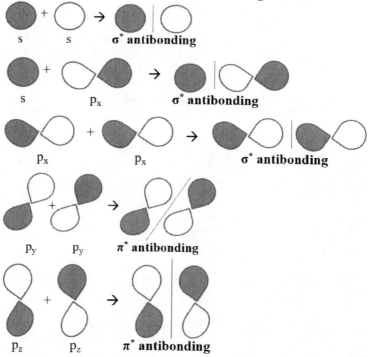

Molecular orbital theory (MOT) – Non-bonding situations

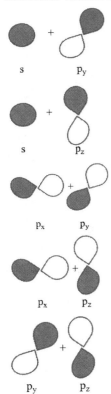

Evidence of delocalization of electrons and resonance structures (bond order)

At Standard Level, it was shown that molecules or ions with double bonds can have several equivalent Lewis structures. In the three resonance structures of the nitrate ion below, the N=O bond should be shorter than the N-O bonds. Experimental measurements show that all three bonds have the same length. This can be shown as the **resonance hybrid** structure and can be explained using the concept of **bond orders**.

The bond lengths and strengths in a delocalized system:
- are equal to one another.
- have values which are intermediate between those of single and double bonds.

$$\text{Bond order} = \frac{\text{total no. of bonding pairs}}{\text{total no. of bonding positions}}$$

Nitrate ion, NO_3^-

[Three resonance structures of nitrate ion showing N with + formal charge, one N=O double bond and two N-O single bonds in each structure, with the double bond in a different position]

[Resonance hybrid structure of NO_3^- showing 120° bond angle]

In NO_3^-, bond order = 4/3 = 1.33

Nitrite ion, NO_2^-

[Two resonance structures of nitrite ion: O=N-O: ↔ :O-N=O]

[Resonance hybrid structure of NO_2^-]

Bond order = 3/2 = 1.5

Sulfite ion, SO_3^{2-}

$$:\ddot{\underset{..}{O}} - S = \ddot{\underset{..}{O}}: \quad \longleftrightarrow \quad :\ddot{\underset{..}{O}} - \ddot{S} - \ddot{\underset{..}{O}}: \quad \longleftrightarrow \quad \ddot{\underset{..}{O}} = \ddot{S} - \ddot{\underset{..}{O}}:$$

$$\left[\ddot{\underset{..}{O}} = \ddot{S} = \ddot{\underset{..}{O}} \right]^{2-}$$

Bond order = 4/3 = 1.33

Carbonate ion, CO_3^{2-}

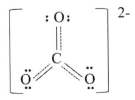

$$\left[\text{structure} \right]^{2-}$$

Bond order = 4/3 = 1.33

Ozone, O_3

Bond order = 3/2 = 1.5

Methanoate ion, HCOO⁻

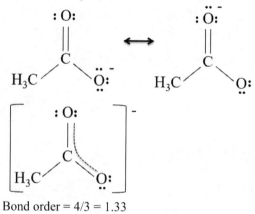

Bond order = 4/3 = 1.33

Ethanoate ion, CH₃COO⁻

Bond order = 4/3 = 1.33

14.2 Hybridization
Syllabus
Nature of science – can you relate this topic to these concepts?
The need to regard theories as uncertain – hybridization in valence bond theory can help explain molecular geometries, but is limited. Quantum mechanics involves several theories explaining the same phenomena, depending on specific requirements.
Understandings – how well can you explain these statements?
A hybrid orbital results from the mixing of different types of atomic orbitals on the same atom.
Applications and skills – how well can you do all of the following?
Explanation of the formation of sp^3, sp^2 and sp hybrid orbitals in methane, ethene and ethyne.
Identification and explanation of the relationships between Lewis (electron dot) structures, electron domains, molecular geometries and types of hybridization.

sp³ hybridization – methane (CH₄)
Carbon has the electronic configuration [He] $2s^2\ 2p_x^1\ 2p_y^1\ 2p_z^0$
According to this theory, it should only form 2 bonds but carbon forms **4 covalent bonds**.

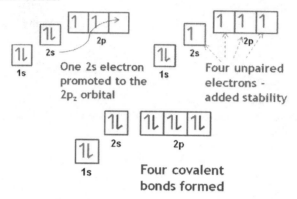

The energy required for this excitation is more than compensated by the formation of four covalent bonds.

In methane (CH₄), all four molecular orbitals are equivalent in shape and energy. They have 25% spherical s character and 75% dumbbell-shaped p character. This allows for stronger bonds to be formed because of the greater overlap.

Molecules with four electron domains will have a tetrahedral electron domain geometry and be predicted to have sp³ hybridization.
If all electron domains are equivalent bonding pair electrons, the molecular geometry will be tetrahedral with bond angles of 109.5°. All bonds between the central atom and the surround atoms are σ bonding.

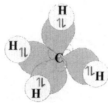

sp^2 hybridization – ethene (C_2H_4)

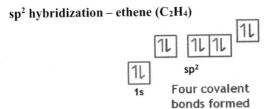

Four covalent bonds formed

The hybridization of one s-orbital and two p-orbitals produces three sp^2 hybrid orbitals which are of equal energies arranged in a trigonal planar geometry. Each sp^2 orbital has 33.3% s character and 66.7% p character. The theoretical bond angle between them is 120°C.
The p_z orbital is unhybridized and are perpendicular to the plane formed by the three sp^2 orbitals. It is at a slightly higher (excited) energy level.

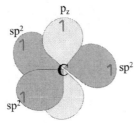

In ethene (C_2H_4), the three sp^2 orbitals are arranged in a trigonal planar shape - each hybrid sp^2 orbital contains one electron and is available to bond with another atom.
- Two sp^2 orbitals from each of the carbon atoms overlap with an s-orbital from four hydrogen atoms to form C-H σ-bonds.
- The third sp^2 orbitals from each of the carbon atoms overlap with each other to form a C-C σ-bond.
- The electrons in the unhybridized p_z-orbital overlap above and below the C-C bond axis to form a C-C π-bond.

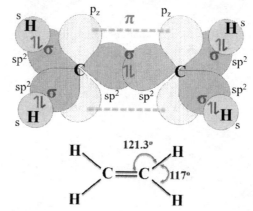

The increased repulsion from the C=C double bond and the extra space it occupies distorts the bond angles as above.

sp hybridization – ethyne (C₂H₂)

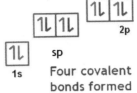

Four covalent bonds formed

The hybridization of one s-orbital and one p-orbitals produces two sp hybrid orbitals which are of equal energies. Each sp orbital has 50% s character and 50% p character. The theoretical bond angle between them is 180°C.

The p_y and p_z orbitals are unhybridized and occupy the remaining Cartesian axes. They are at a slightly higher (excited) energy level.

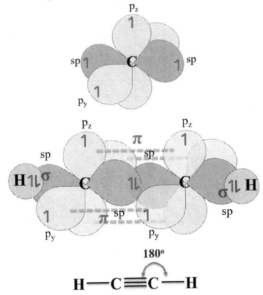

H—C≡C—H

The two sp orbitals are arranged in a linear shape - each hybrid sp orbital contains one electron and is available to bond with another atom.

- One sp orbital from each of the carbon atoms overlaps with an s-orbital from two hydrogen atoms to form C-H σ-bonds.
- The second sp orbitals from each of the carbon atoms overlap with each other to form a C-C σ-bond.
- The electrons in the unhybridized p-orbitals, orientated at 90° to each other overlap as a cylinder around the C-C bond axis to form two C-C π-bonds.

Hybridization and molecular geometry

Hybridization can be used to predict the shapes of molecules and vice versa. The relationships between electron domain geometry and hybridization are:
- tetrahedral – sp^3 hybridization
- trigonal planar – sp^2 hybridization
- linear – sp hybridization

If all the hybridized orbitals are bonding pair electrons, the molecular geometry will be the same as the electron domain geometry. The repulsion of lone pairs of electrons determines the shape of the molecules but they are not considered when describing the shape of the molecule.

CO_2

O = C = O

No. of electron domains: 2
Geometrical arrangement of electron domains: linear
Shape of molecule: linear
Hybridization: sp

BF_3

No. of electron domains: 3
Geometrical arrangement of electron domains: trigonal planar
Shape of molecule: trigonal planar
Hybridization: sp^2

SO_2

No. of electron domains: 3
Geometrical arrangement of electron domains: trigonal planar
Shape of molecule: V-shaped
Hybridization: sp^2

CH₄

No. of electron domains: 4
Geometrical arrangement of electron domains: tetrahedral
Shape of molecule: tetrahedral
Hybridization: sp³

NH₃

No. of electron domains: 4
Geometrical arrangement of electron domains: tetrahedral
Shape of molecule: trigonal pyramidal
Hybridization: sp³

H₂O

No. of electron domains: 4
Geometrical arrangement of electron domains: tetrahedral
Shape of molecule: V-shaped
Hybridization: sp³

5.1 Measuring energy changes
Syllabus
Nature of science – can you relate this topic to these concepts?
Fundamental principle – conservation of energy is a fundamental principle of science.
Making careful observations – measurable energy transfers between systems and surroundings.
Understandings – how well can you explain these statements?
Heat is a form of energy.
Temperature is a measure of the average kinetic energy of the particles.
Total energy is conserved in chemical reactions.
Chemical reactions that involve transfer of heat between the system and the surroundings are described as endothermic or exothermic.
The enthalpy change (ΔH) for chemical reactions is indicated in kJ mol^{-1}.
ΔH values are usually expressed under standard conditions, given by $\Delta H^\ominus$, including standard states.
Applications and skills – how well can you do all of the following?
Calculation of the heat change when the temperature of a pure substance is changed using $q = mc\Delta T$.
A calorimetry experiment for an enthalpy of reaction should be covered and the results evaluated.

Thermal energy (heat) transfer in chemical reactions

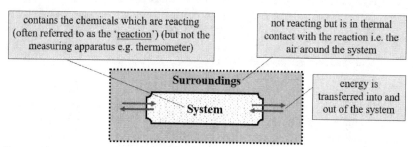

Open system
Energy and matter can be exchanged between the system and the surroundings.

Closed system
Energy can be exchanged between the system and the surroundings, but not matter.

Thermal energy (heat, q)
Chemical energy in the system at constant pressure that can be converted into heat upon reaction.

Temperature, T is a measure of the **average kinetic energy** of the particles. The absolute temperature (in kelvin, K) is proportional to the average kinetic energy of the particles.

At **absolute zero (0 K)**, particles stop moving. The **entropy, S** which is a measure of the disorder of the particles reaches a minimum value.

Enthalpy and enthalpy change

The total **enthalpy** of a system cannot be measured directly. The **enthalpy change (ΔH)** of a system is measured instead. Therefore, enthalpy is a **state function** as it depends on the initial and final state of the system.

The enthalpy change in a system can be calculated by measuring its effect on the surroundings (temperature change). This is possible because of the **first law of thermodynamics** or the **law of conservation of energy** which states that energy can neither be created of destroyed (only converted into different forms). This is the amount of energy (in kJ) released or absorbed when a chemical reaction occurs under standard conditions. As with most chemical quantities, it is specified per mole of substance as in the balanced chemical equation for the reaction.

Standard conditions for enthalpy change:
- A temperature of 298 K (25°C)
- A pressure of 101 kPa or 1 atm
- A concentration of 1.00 mol dm^{-3} for solutions.
- Reactants and products in physical states, normal for the above conditions e.g. Br_2 should be liquid, NaCl should be solid, O_2 should be gas, C should be graphite (not diamond), etc.

Standard enthalpy change of reaction

ΔH: **the amount of energy (in kJ) released or absorbed when a chemical reaction occurs under standard conditions.**

As with most chemical quantities, it is specified per mole of substance as in the balanced chemical equation for the reaction.

$\Delta H = H_{products} - H_{reactants}$

Unit for ΔH: kJ mol^{-1}

Standard conditions for enthalpy change
- A **temperature** of 298K (25°C)
- A **pressure** of 100 kilopascals (100 kPa / 1 atm)
- A **concentration** of 1.0 mol dm^{-3} for solutions.
- Reactants and products in **physical states, normal** for the above conditions e.g. Br_2 should be liquid, NaCl should be solid, O_2 should be gas, C should be graphite (not diamond), etc.

Endothermic and exothermic reactions
Exothermic reaction: Chemical energy is converted into heat energy and the temperature of the system **rises**.
The system loses energy which eventually goes into the surrounding as heat. The surrounding feels warmer.
Endothermic reaction: Heat energy is converted into chemical energy and the temperature of the system **falls**.
The system absorbs heat energy from the surrounding. The surrounding feels cooler.

Enthalpy level diagrams
Exothermic
$\Delta H = H_{products} - H_{reactants} = -ve$

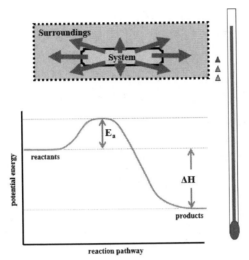

Endothermic
$\Delta H = H_{products} - H_{reactants} = +ve$

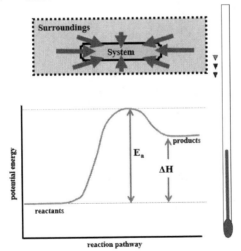

These diagrams are called enthalpy level diagrams, energy level diagrams or reaction profiles (according to which textbook you're using!).

Exothermic Reactions (examples)
Zinc metal displaces copper from a copper sulfate solution.
$Zn\,(s) + CuSO_4\,(aq) \rightarrow ZnSO_4\,(aq) + Cu\,(s)$
$\Delta H = -85\text{ kJ mol}^{-1}$
Combustion of methane in air (oxygen).
$CH_4\,(g) + 2O_2\,(g) \rightarrow CO_2\,(g) + 2H_2O\,(l)$
$\Delta H = -890\text{ kJ mol}^{-1}$

Endothermic Reactions (examples)
Thermal decomposition of calcium carbonate.
$CaCO_3\,(s) \rightarrow CaO\,(s) + CO_2\,(g)$
$\Delta H = +100\text{ kJ mol}^{-1}$
Dissolving ammonium chloride.
$NH_4Cl\,(s) \rightarrow NH_4^+\,(aq) + Cl^-\,(aq)$
$\Delta H = +12\text{ kJ mol}^{-1}$

Standard enthalpy changes (Definitions)

Standard enthalpy change of reaction $\Delta H°_r$
The enthalpy change when the amounts of reactants shown in the equation for the reaction, react under standard conditions to give the products in their standard states.

Standard enthalpy change of formation $\Delta H°_f$
The enthalpy change when one mole of a compound is formed from its elements under standard conditions; both compound and elements are in standard states.

Standard enthalpy change of combustion $\Delta H°_c$
The enthalpy change when one mole of an element or compound reacts completely with oxygen under standard conditions.

These are defined in terms of the **reactant(s)**:

Standard enthalpy change of reaction $\Delta H°_r$
The enthalpy change when the amounts of reactants shown in the equation for the reaction, react at 298K and 1 atm to give the products in their standard states.
$2Na\ (s) + Cl_2\ (g) \rightarrow 2NaCl\ (s)$
$\Delta H°_r = -822\ kJ$
$Na\ (s) + \frac{1}{2}Cl_2\ (g) \rightarrow NaCl\ (s)$
$\Delta H°_r = -411\ kJ$

Standard enthalpy change of combustion $\Delta H°_c$
The enthalpy change when one mole of an element or compound reacts completely with oxygen at a stated temperature (usually 298K) and 1 atm.
$C_2H_5OH\ (l) + 3O_2\ (g) \rightarrow 2CO_2\ (g) + 3H_2O\ (l)$
$\Delta H°_c = -1371\ kJ\ mol^{-1}$

These are defined in terms of the **product**:

Standard enthalpy change of neutralisation $\Delta H°_{neut}$
The enthalpy change when an acid is neutralised by an alkali to produce 1 mole of water, at 298K and 1 atm; all solutions have a concentration of 1 mol dm^{-3}.
$H^+\ (aq) + OH^-\ (aq) \rightarrow H_2O\ (l)$
$HCl\ (aq) + NaOH\ (aq) \rightarrow NaCl\ (aq) + H_2O\ (l)$
$\Delta H°_{neut} = -57.9\ kJ\ mol^{-1}$
$\frac{1}{2}H_2SO_4\ (aq) + NaOH\ (aq) \rightarrow \frac{1}{2}Na_2SO_4\ (aq) + H_2O\ (l)$
$\Delta H°_{neut} = -57.9\ kJ\ mol^{-1}$

Standard enthalpy change of formation $\Delta H°_f$
The enthalpy change when one mole of a substance is formed from its elements at 298K and 1 atm; both compound and elements are in standard states.
$2C\ (s) + 3H_2\ (g) + \frac{1}{2}O_2\ (g) \rightarrow C_2H_5OH\ (l)$
$\Delta H°_f = -277.1\ kJ\ mol^{-1}$

Standard enthalpy change of atomisation $\Delta H°_a$
The enthalpy change when one mole of gaseous atoms is formed from an element in its standard state at 298K and 1 atm.
$\frac{1}{2}I_2\ (s) \rightarrow I\ (g)$
$\Delta H°_a = +106.1\ kJ\ mol^{-1}$
$Na\ (s) \rightarrow Na\ (g)$
$\Delta H°_a = +97.7\ kJ\ mol^{-1}$

Thermochemistry experiment – combustion of a fuel
Example: **Combustion of ethanol.**
$C_2H_5OH\ (l) + 3O_2\ (g) \rightarrow 2CO_2\ (g) + 3H_2O\ (l)$

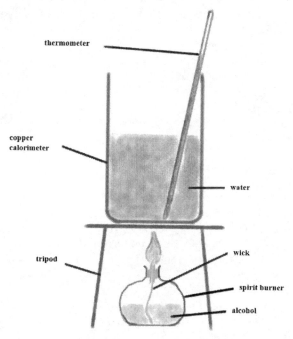

Procedure:
1. The spirit burner containing ethanol is weighed.
2. A known volume of water e.g. 100cm³ is poured into the copper calorimeter. (Copper is used as it is a very good conductor of heat).
3. The temperature of the water is taken at regular intervals e.g. 1 minute for about 5 minutes.
4. The burner is lit and the temperature is recorded regularly.
5. Extinguish the burner after the temperature has reached about 20°C above room temperature.
6. The temperature continues to be recorded for a further 5 minutes.
7. The mass of the spirit burner is taken.

Analysis:
A plot of temperature against time is made.

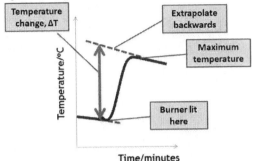

Energy change = m x c x ΔT
m = mass of water (1 cm³ = 1g)
c = specific heat capacity (of water) = 4.18 kJ kg⁻¹ K⁻¹
Mass of ethanol burnt = initial mass of burner – final mass of burner
Amount of ethanol burnt = mass of ethanol burnt / molar mass of ethanol
Enthalpy change of combustion, ΔH_c = q / moles (in kJ mol⁻¹)

Sources of error:
1. Heat is lost to the surrounding as the experiment takes a long time. Extrapolating the graph for ΔT allows for only some of this heat loss.
2. Some heat is absorbed by the air and not the water.
3. Some heat is absorbed by the copper calorimeter and not the water.
4. Some of the ethanol may not burn completely or at all. Incomplete combustion produces black soot on the bottom of the beaker. Unburnt ethanol evaporated into the air.
5. Conditions are not standard for the reaction. As the reaction is very exothermic, water is produced as steam and not liquid.

These errors can be reduced by using a flame calorimeter:
Sometimes the specific heat capacity of water is used, but in more accurate flame calorimeters, the actual heat capacity of the flame calorimeter can be determined and used (by finding the temperature rise when a known amount of a substance with an accurately known enthalpy of combustion is tested).

Thermochemistry experiment – reaction in aqueous solutions
Example: **Displacement reaction between iron and copper sulfate solution**

Fe (s) + CuSO$_4$ (aq) → FeSO$_4$ (aq) + Cu (s)

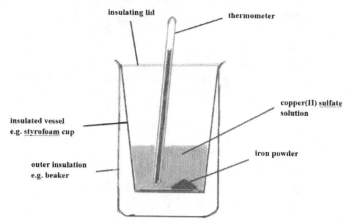

Procedure:
1. A measured volume of copper (II) sulfate solution of known concentration is pipetted into an insulated container (polystyrene cup).
2. An excess (slightly) of iron powder is weighed out.
3. The temperature of the solution is taken at regular intervals e.g. 30 seconds for about 2 minutes.
4. The metal powder is added and the reaction mixture is stirred.
5. The temperature is recorded for a further 2 minutes after the maximum temperature is reached.

Assumptions:
- The density of the solution is the same as water, which is 1 g cm^{-3}.
- The mass of the metal powder is not taken into account.

Metal powder is used so that the reaction proceeds as fast as possible.

Analysis:
A plot of temperature against time is made.

Amount of $CuSO_4$ = conc. x volume
Change the unit for volume to dm^3 as the concentration is quoted in mol dm^{-3}

Energy change = m x c x ΔT
m = mass of solution (1 cm^3 = 1g)
c = specific heat capacity (of water) = 4.18 kJ kg^{-1} K^{-1}
Enthalpy change of reaction, ΔH_r = q / moles (in kJ mol^{-1})

Sources of error:
1. Heat is lost to the surrounding as the reaction is relatively slow. Extrapolating the graph for ΔT allows for most of this heat loss.
2. Some heat is absorbed by the metal and thermometer.

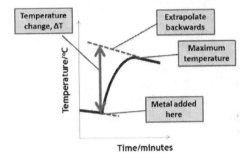

Thermochemistry experiment – neutralisation
Example: **Neutralisation between hydrochloric acid and sodium hydroxide**

HCl (aq) + NaOH (aq) → NaCl (aq) + H$_2$O (l)

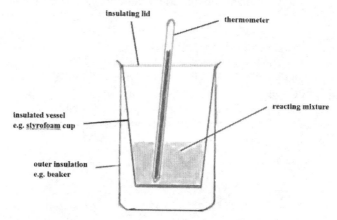

Procedure:
1. A measured volume of dilute hydrochloric acid of known concentration is pipetted into an insulated container (polystyrene cup). The concentration should be at least 2.00 mol dm^{-3} to give a reasonable temperature rise.
2. An exact volume of sodium hydroxide solution enough to neutralise the acid is measured out.
3. The temperature of the acid is taken at regular intervals e.g. 30 seconds for about 2 minutes.
4. The alkali is added and the reaction mixture is stirred.
5. The temperature is recorded for a further 2 minutes after the maximum temperature is reached.

Assumption:
- The densities of the solutions are the same as water, which is 1 g cm^{-3}.

Analysis:
A plot of temperature against time is made.

Amount of HCl = conc. x volume
Change the unit for volume to dm^3 as the concentration is quoted in mol dm^{-3}

Energy change = m x c x ΔT
m = mass of solution (1 cm^3 = 1g)
c = specific heat capacity (of water) = 4.18 kJ kg^{-1} K^{-1}
Enthalpy change of neutralisation, ΔH_{neut} = q / moles (in kJ mol^{-1})

Sources of error:
1. Heat is lost to the surrounding as the reaction is relatively slow. Extrapolating the graph for ΔT allows for most of this heat loss.
2. Some heat is absorbed by the stirrer and thermometer.

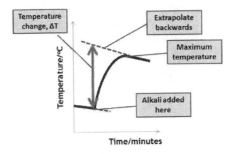

Calculating ΔH_r from ΔH_f

Enthalpies of reaction, ΔH_r can be calculated using enthalpies of formation, ΔH_f data using the following equation:

$\Delta H_r = \Sigma \Delta H_f$ (products) $- \Sigma \Delta H_f$ (reactants)

The values for ΔH_f can be obtained from Table 12 of the Data Booklet. The enthalpy change of formation of elements in their normal states under standard conditions is defined as zero.

Example calculations:

Hydrogenation of ethene
C_2H_4 (g) + H_2 (g) $\rightarrow$ C_2H_6 (g)
$\Delta H_r = [-84.0 - (+52.0) - 0]$ kJ $= -136.0$ kJ

Combustion of cyclohexane
C_6H_{12} (l) + $9O_2$ (g) $\rightarrow$ $6CO_2$ (g) + $6H_2O$ (l)
$\Delta H_r = [6 \times (-393.5) + 6 \times (-285.8) - (-156) - 9 \times 0]$ kJ $= -2919.8$ kJ

5.2 Hess's law
Syllabus
Nature of science – can you relate this topic to these concepts?
Hypotheses – based on the conservation of energy and atomic theory, scientists can test the hypothesis that if the same products are formed from the same initial reactants then the energy change should be the same regardless of the number of steps.
Understandings – how well can you explain these statements?
The enthalpy change for a reaction that is carried out in a series of steps is equal to the sum of the enthalpy changes for the individual steps.
Applications and skills – how well can you do all of the following?
Application of Hess's Law to calculate enthalpy changes.
Calculation of ΔH reactions using ΔH_f data.
Determination of the enthalpy change of a reaction that is the sum of multiple reactions with known enthalpy changes.

Hess's law
Hess's law states that if the initial and final states of the system are the same, the enthalpy change of a chemical reaction will be the same regardless of the route taken.
The enthalpy change for a reaction that is carried out in a series of steps is equal to the sum of the enthalpy changes for the individual steps.
Consider this reaction: A + B → C + D
Suppose that the production of C and D could be achieved by going through an extra step involving the production of X and Y.

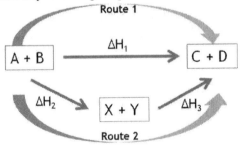

According to Hess's law,
$\Delta H_1 = \Delta H_2 + \Delta H_3$
i.e. the enthalpy change following route 1 is exactly the same as the (total) enthalpy following route 2.

Questions involving Hess's law usually involve a cycle similar to this:

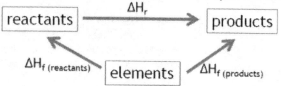

ΔH_r could be any reaction e.g. neutralisation, displacement, combustion, etc.
ΔH_f is the enthalpy of formation.

Example 1 (ΔH_c from ΔH_f):

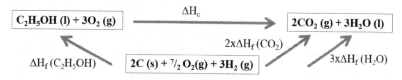

$\Delta H_c = 2 \times \Delta H_f (CO_2) + 3 \times \Delta H_f (H_2O) - \Delta H_f (C_2H_5OH)$
There is no need to calculate ΔH_f for O_2 by definition.
The elements are written once, but 'used' twice in the cycle.

Example 2 (ΔH_r from ΔH_c):

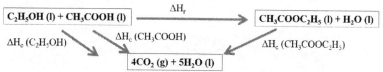

$\Delta H_r = \Delta H_c (C_2H_5OH) + \Delta H_c (CH_3COOH) - \Delta H_c (CH_3COOC_2H_5)$
Water does not 'burn'.
Complete combustion is assumed for all cases. There is no need to add extra O_2 (g) to the equations or cycle.

Practice questions
The answers to these questions have been given deliberately for checking. It is more important that you practise constructing Hess's cycles for various reactions.

Question 1: Calculate the standard enthalpy of combustion for ethane.
Use the following data.
$\Delta H°_f (C_2H_6) = -84.0$ kJ mol^{-1}
$\Delta H°_f (CO_2) = -393.5$ kJ mol^{-1}
$\Delta H°_f (H_2O\ (l)) = -285.8$ kJ mol^{-1}
Answer = -1560.4 kJ mol^{-1}

Question 2: Calculate the standard enthalpy of combustion for octane.
Use the following data.
$\Delta H°_f (C_8H_{18}) = -210.0$ kJ mol^{-1}
$\Delta H°_f (CO_2) = -393.5$ kJ mol^{-1}
$\Delta H°_f (H_2O\ (l)) = -285.8$ kJ mol^{-1}
Answer = -5507.5 kJ mol^{-1}

Question 3: Calculate the standard enthalpy of combustion for ethanol.
Use the following data.
$\Delta H°_f (C_2H_5OH) = -278$ kJ mol^{-1}
$\Delta H°_f (CO_2) = -393.5$ kJ mol^{-1}
$\Delta H°_f (H_2O\ (l)) = -285.8$ kJ mol^{-1}
Answer = -1366.4 kJ mol^{-1}

Question 4: Calculate the standard enthalpy of formation for propane.
Use the following data.
$\Delta H°_c$ (C_3H_8) = -2219 kJ mol^{-1}
$\Delta H°_c$ (C) = -394 kJ mol^{-1}
$\Delta H°_c$ (H_2) = -286 kJ mol^{-1}
Answer = -107 kJ mol^{-1}

Question 5: Calculate the standard enthalpy of formation for ethanol.
Use the following data.
$\Delta H°_c$ (C_2H_5OH) = -1367 kJ mol^{-1}
$\Delta H°_c$ (C) = -394 kJ mol^{-1}
$\Delta H°_c$ (H_2) = -286 kJ mol^{-1}
Answer = -279 kJ mol^{-1}

Question 6: Calculate the standard enthalpy of formation for glucose.
Use the following data.
$\Delta H°_c$ ($C_6H_{12}O_6$) = -2803 kJ mol^{-1}
$\Delta H°_c$ (C) = -394 kJ mol^{-1}
$\Delta H°_c$ (H_2) = -286 kJ mol^{-1}
Answer = -1277 kJ mol^{-1}

5.3 Bond enthalpy
Syllabus
Nature of science – can you relate this topic to these concepts?
Models and theories – measured energy changes can be explained based on the model of bonds broken and bonds formed. Since these explanations are based on a model, agreement with empirical data depends on the sophistication of the model and data obtained can be used to modify theories where appropriate.
Understandings – how well can you explain these statements?
Bond-forming releases energy and bond-breaking requires energy. Average bond enthalpy is the energy needed to break one mol of a bond in a gaseous molecule averaged over similar compounds.
Applications and skills – how well can you do all of the following?
Calculation of the enthalpy changes from known bond enthalpy values and comparison of these to experimentally measured values.
Sketching and evaluation of potential energy profiles in determining whether reactants or products are more stable and if the reaction is exothermic or endothermic.
Discussion of the bond strength in ozone relative to oxygen in its importance to the atmosphere.

Bond enthalpy (bond dissociation enthalpy), BE
The energy required to break 1 mol of bonds in gaseous covalent molecules under standard conditions. The precise energy of a bond is subject to variation due to the effects of neighbouring atoms or groups: values are always quoted as an average determined over a wide variety of molecules e.g.

H-H (g) → 2H (g)   BE = +436 kJ mol^{-1}
O=O (g) → 2O (g)   BE = +498 kJ mol^{-1}
Cl-Cl (g) → 2Cl (g)   BE = +242 kJ mol^{-1}

Conversely, the energy released when 1 mole of a particular type of covalent bond is formed has the same numerical value but is negative (exothermic).
2H(g) → H-H (g)   BE = -436 kJ mol^{-1}
The values in Table 11 of the Data Booklet are given as positive.

Bond enthalpy and bond length
Covalent bonds are the result of two atoms sharing pairs of electrons. The atoms are held together by the mutual attraction between the nuclei and the shared electrons.
The strength of the bond(s) are dependent on the length of the bond(s) and the radii of the atoms.
The bond will be weaker if:
- the atomic radii are larger.
- the bond length is longer.

Bond	H-F	H-Cl	H-Br	H-I
Bond length / pm	92	128	141	160
Bond enthalpy / kJ mol^{-1}	567	431	366	151

Bond enthalpy and multiple bonds
Bond enthalpy increases with multiple bonds.

Bond	O-O	O=O	C-C	C=C	C≡C	N-N	N=N	N≡N
Bond enthalpy / kJ mol^{-1}	144	498	346	614	839	158	470	945

Bond enthalpy and bond polarity
An increase in the electronegativity difference between the atoms leads to an increase in ionic character and bond polarity. This causes the bond to be stronger (bond enthalpy increases).

Bond	H-F	H-Cl	H-Br	H-I
$\Delta\chi$	1.8	1.0	0.8	0.5
Bond enthalpy / kJ mol^{-1}	567	431	366	151

Calculating Energy Changes
Enthalpy change of a reaction can be calculated using bond enthalpy data as follows:
ΔH° = Σ (BE bonds broken) – Σ (BE bonds formed)
Note: it is helpful when doing these to draw the full structural formula of any relevant molecules so as to count the number of bonds being broken and formed accurately.

Example:
N_2 (g) + $2O_2$ (g) → $2NO_2$ (g)
Showing all the bonds: N≡N + 2 O=O → 2 O=N-O
ΔH = [$BE_{N≡N}$ + $2BE_{O=O}$] - [$2BE_{O=N}$ + $2BE_{O-N}$] = [(2×587)+(2×214)]-[945+(2×498)]
ΔH = -339 kJ mol^{-1}

Exercise:
Calculate the enthalpy change for the following reactions using bond enthalpy data. The experimental values for ΔH have been given in each case. How do your values differ to them?

1. $2H_2$ (g) + O_2 (g) → $2H_2O$ (g) ΔH = -241.8 kJ mol^{-1}
2. Cl_2 (g) + F_2 (g) → 2ClF (g) ΔH = -108.1 kJ mol^{-1}
3. C_2H_6 (g) + Cl_2 → C_2H_5Cl (g) + HCl (g) ΔH = -122.3 kJ mol^{-1}
4. C_2H_2 (g) + $2H_2$ (g) → C_2H_6 (g) ΔH = -296.5 kJ mol^{-1}
5. C_2H_4 (g) + H_2O (g) → C_2H_5OH (g) ΔH = -44.9 kJ mol^{-1}
6. CH_4 (g) + $2O_2$ → CO_2 (g) + $2H_2O$ (l) ΔH = -891 kJ mol^{-1}
    Note that: H_2O (l) → H_2O (g) ΔH = +41 kJ mol^{-1}

Ozone (Trioxygen)
Ozone, O_3 exists in the stratosphere of the Earth's atmosphere and absorbs harmful UV radiation from the Sun, preventing it from damaging cells in plants and animals. Under normal circumstances it is created and destroyed at comparable rates, where an equilibrium is reached.

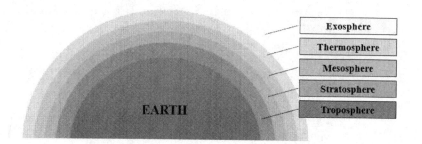

Formation of Ozone
Step 1: An oxygen molecule absorbs UV radiation and splits into two oxygen atoms.
$O_2 (g) \rightarrow O\bullet (g) + O\bullet (g)$
$\Delta H = + 498$ kJ mol^{-1}
Step 2: The oxygen atom attacks another oxygen molecule to form ozone.
$O\bullet (g) + O_2 (g) \rightarrow O_3 (g)$

(Natural) Destruction of Ozone
One of the oxygen-oxygen bonds is broken by UV radiation to form an oxygen molecule and an oxygen atom.
$O_3 (g) \rightarrow O\bullet (g) + O_2 (g)$
$\Delta H = + 364$ kJ mol^{-1}

(Anthropogenic) Destruction of Ozone
The excessive use of chlorofluorocarbons (CFCs) e.g. Freon as refrigerants, aerosol propellants and solvents led to large scale depletion of the ozone layer.
In the stratosphere, chlorine (atomic) gas was released when CFCs were exposed to UV radiation.
The chlorine atoms (Cl•) attacked the ozone molecules while releasing chlorate free radicals (ClO•) which further attacked other ozone molecules.
$Cl\bullet (g) + O_3 (g) \rightarrow ClO\bullet (g) + O_2 (g)$
This reaction in turn, released another chlorine atom and further propagated this chain reaction.
$ClO\bullet (g) + O_3 (g) \rightarrow Cl\bullet (g) + 2O_2 (g)$
Thus, a small amount of CFCs could destroy enormous quantities of ozone.

15.1 Energy cycles
Syllabus
Nature of science – can you relate this topic to these concepts?
Making quantitative measurements with replicates to ensure reliability—energy cycles allow for the calculation of values that cannot be determined directly.
Understandings – how well can you explain these statements?
Representative equations (e.g. M^+ (g) → M^+ (aq)) can be used for enthalpy/energy of hydration, ionization, atomization, electron affinity, lattice, covalent bond and solution.
Enthalpy of solution, hydration enthalpy and lattice enthalpy are related in an energy cycle.
Applications and skills – how well can you do all of the following?
Construction of Born-Haber cycles for group 1 and 2 oxides and chlorides.
Construction of energy cycles from hydration, lattice and solution enthalpy. For example dissolution of solid NaOH or NH_4Cl in water.
Calculation of enthalpy changes from Born-Haber or dissolution energy cycles.
Relate size and charge of ions to lattice and hydration enthalpies.
Perform lab experiments which could include single replacement reactions in aqueous solutions.

Standard states
Thermochemical equations show the substances in their **standard states**.
e.g. Br_2 (l), NaCl (s), O_2 (g), C (s) is graphite (not diamond), etc.
Always include the state symbols when writing thermochemical equations.

Standard conditions
Most enthalpy changes are quoted as **standard enthalpy changes** under **standard conditions** which are:
- A **temperature** of 298K (25°C), not usually shown but assumed.
- A **pressure** of 1 atm.
- A **concentration** of 1.0 mol dm^{-3} for solutions (this is not usually stated).

Standard conditions are denoted by the superscript $^\ominus$, e.g. $\Delta H^\ominus$

Standard enthalpy changes

Standard enthalpy change of formation $\Delta H°_f$

The enthalpy change when **one mole** of a substance is **formed** from its elements at **298K** and **1 atm**; both compound and elements are in **standard states**.

$2C (s) + 3H_2 (g) + ½O_2 (g) \rightarrow C_2H_5OH (l)$
$\Delta H°_f = -277.1$ kJ mol^{-1}

Lattice enthalpy (ΔH_{lat})

Energy change when **gaseous ions** are formed from **1 mol** of a **solid ionic compound** in its **lattice**.

$MX (s) \rightarrow M^+ (g) + X^- (g)$ ΔH_{lat} positive

Lattice enthalpy cannot be measured directly. It can be calculated from experimental values of other measurable enthalpies in a Born-Haber cycle.

Alternative definition:
Energy change when 1 mol of a solid ionic compound is formed from its constituent gaseous ions.

$M^+ (g) + X^- (g) \rightarrow MX (s)$ ΔH_{lat} negative.

Enthalpy of atomisation (ΔH_{at})

Energy change when **one mole** of **gaseous atoms** is formed from its element in its **standard state** e.g.

$Na(s) \rightarrow Na (g)$
$½Cl_2 (g) \rightarrow Cl (g)$ not $Cl_2 (g) \rightarrow 2Cl (g)$

Electron affinity (ΔH_{EA})

First electron affinity (ΔH_{EA1}): The energy required to add **one electron** to each **gaseous atom** in **one mole** of an element. This is usually an exothermic process – a negatively charged electron is brought towards a positively charged nucleus.
$A\,(g) + e^- \rightarrow A^-\,(g)$ ΔH_{EA1} (usually negative)

Second electron affinity (ΔH_{EA2}): The energy required to add **one electron** to each singly charged **gaseous negative ion** in **one mole** of an element. Second electron affinity is always an endothermic process – a negatively charged electron is brought towards a negatively charged ion which repels it.
$A^-\,(g) + e^- \rightarrow A^{2-}\,(g)$ ΔH_{EA2} (always positive)

Born-Haber cycles

Example 1: Calculating ΔH_f from a Born-Haber cycle for NaCl

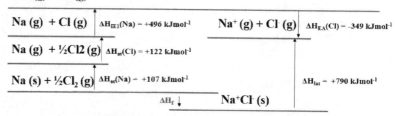

$\Delta H_f = \Delta H_{at}(Na) + \Delta H_{at}(Cl) + \Delta H_{IE1}(Na) + \Delta H_{EA}(Cl) + \Delta H_{lat}$
$\Delta H_f = [107 + 122 + 496 - 349 - 790] = -414$ kJ mol^{-1}

Example 2: Calculating ΔH_f from a Born-Haber cycle for MgCl$_2$.

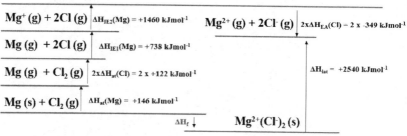

$\Delta H_f = \Delta H_{at}(Mg) + 2\Delta H_{at}(Cl) + \Delta H_{IE1}(Mg) + \Delta H_{IE2}(Mg) + 2\Delta H_{EA}(Cl) + \Delta H_{lat}$
$\Delta H_f = 146 + (2 \times 122) + 738 + 1460 + (2 \times -349) - 2540 = -650$ kJ mol^{-1}

Example 3: Calculating ΔH_{lat} from a Born-Haber cycle for $CaCl_2$.

$Ca^{2+}(g) + 2Cl^-(g)$

$Ca^+(g) + 2Cl(g)$ $\quad \Delta H_{IE2}(Ca) = +1150 \text{ kJmol}^{-1}$	$Ca^{2+}(g) + 2Cl^-(g) \quad 2\times\Delta H_{EA}(Cl) = 2 \times -349 \text{ kJmol}^{-1}$
$Ca(g) + 2Cl(g) \quad \Delta H_{IE1}(Ca) = +590 \text{ kJmol}^{-1}$	
$Ca(g) + Cl_2(g) \quad 2\times\Delta H_{at}(Cl) = 2 \times +122 \text{ kJmol}^{-1}$	ΔH_{lat}
$Ca(s) + Cl_2(g) \quad \Delta H_{at}(Ca) = +178 \text{ kJmol}^{-1}$	
$\Delta H_f = -796 \text{ kJmol}^{-1} \downarrow$	$Ca^{2+}(Cl^-)_2(s)$

$\Delta H_{lat} = -\Delta H_f + \Delta H_{at}(Ca) + 2\Delta H_{at}(Cl) + \Delta H_{IE1}(Ca) + \Delta H_{IE2}(Ca) + 2\Delta H_{EA}(Cl)$

$\Delta H_{lat} = +796 + 178 + (2 \times 122) + 590 + 1150 + (2 \times -349) = 2260 \text{ kJ mol}^{-1}$

Example 4: Calculating ΔH_{lat} from a Born-Haber cycle for CaO.

$Ca^{2+}(g) + O(g)$

$Ca^+(g) + O(g) \quad \Delta H_{IE2}(Ca) = +1150 \text{ kJmol}^{-1}$	$Ca^{2+}(g) + O^-(g) \quad \Delta H_{EA1}(O) = -141 \text{ kJ mol}^{-1}$
$Ca(g) + O(g) \quad \Delta H_{IE1}(Ca) = +590 \text{ kJmol}^{-1}$	$Ca^{2+}(g) + O^{2-}(g) \quad \Delta H_{EA2}(O) = +753 \text{ kJ mol}^{-1}$
$Ca(g) + \frac{1}{2}O_2(g) \quad \Delta H_{at}(O) = +121 \text{ kJmol}^{-1}$	
$Ca(s) + \frac{1}{2}O_2(g) \quad \Delta H_{at}(Ca) = +178 \text{ kJmol}^{-1}$	ΔH_{lat}
$\Delta H_f = -635 \text{ kJmol}^{-1} \downarrow$	$Ca^{2+}O^{2-}(s)$

$\Delta H_{lat} = -\Delta H_f + \Delta H_{at}(Ca) + \Delta H_{at}(O) + \Delta H_{IE1}(Ca) + \Delta H_{IE2}(Ca) + \Delta H_{EA1}(O) + \Delta H_{EA1}(O)$

$\Delta H_{lat} = +635 + 178 + 121 + 590 + 1150 - 141 + 753 = -3286 \text{ kJ mol}^{-1}$

Lattice enthalpy trends – ionic radii

Lattice enthalpy decreases as the size of the ion increases. This applies to both cations and anions. The reason is that as the radius of the ion increases, the electrostatic attraction between the ions decreases so the lattice enthalpy is smaller.

Cation	Anion	ΔH_{lat} (kJ mol^{-1})			
		F$^-$	Cl$^-$	Br$^-$	I$^-$
	Ionic radius (pm)	*133*	*181*	*196*	*220*
Li$^+$	*76*	1049	864	820	764
Na$^+$	*102*	930	790	754	705
K$^+$	*138*	829	720	691	650
Rb$^+$	*152*	795	695	668	632
Cs$^+$	*167*	759	670	647	613

Lattice enthalpy trends – ionic charge

Lattice enthalpy increases as the charge on the ion increases. This applies to both cations and anions. The reason is that as the charge on the ion increases, the electrostatic attraction between the ions increases so more energy is required to break the lattice.

		ΔH_{lat} (kJ mol^{-1})	
	Anion	Cl$^-$	O^{2-}
Cation	*Ionic charge*	*1-*	*2-*
Li$^+$	*1+*	864	
Be^{2+}	*2+*	3033	
Ba^{2+}	*2+*	2069	3054

A clear trend can be seen by comparing NaF with MgO: both have the same electronic structure.
- the Na$^+$ ion is about the same size as the Mg^{2+} ion.
- the F$^-$ ion is about the same size as the O^{2-} ion.

The only difference between the two solids is the *charge* on the ions.

NaF ΔH_{lat} = -930 kJ mol^{-1}
MgO ΔH_{lat} = -3791 kJ mol^{-1}

The lattice enthalpy for magnesium oxide is much more negative, which shows that doubly charged ions attract each other more strongly than singly charged ions.

Solutions

Standard enthalpy change of solution (ΔH_{sol})

Enthalpy change when one mole of a substance is dissolved in sufficient solvent to give an infinitely dilute solution e.g.
NH_4NO_3 (s) → NH_4^+ (aq) + Cl^- (aq)　　ΔH_{sol} = +25.69 kJ mol^{-1}
KCl (s) → K^+ (aq) + Cl^- (aq)　　ΔH_{sol} = +17.22 kJ mol^{-1}

Enthalpy change of hydration (ΔH_{hyd})

Enthalpy change when one mole of gaseous ions are dissolved in sufficient water to give an infinitely dilute solution. The enthalpy of hydration for the cation and anion has to be considered separately e.g.
Mg^{2+} (g) → Mg^{2+} (aq)　　ΔH_{hyd} = -1963 kJ mol^{-1}
NO_3^- (g) → NO_3^- (aq)　　ΔH_{hyd} = -316 kJ mol^{-1}
For solvents other than water, the term solvation is used instead of hydration.
An infinitely dilute solution is one in which further dilution does not cause an enthalpy change.

Water molecules are **polar** (they have **permanent dipoles**). This is due to the difference in electronegativities between the oxygen atom and the hydrogen atoms. The two lone pairs of electrons repel the electrons in the covalent O-H bond and H_2O has a bent (V-shaped) geometry.

Dissolution (dissolving in solvent) can be thought of to consist of two processes:
1. Breaking of the ionic lattice which involves the **lattice enthalpy, ΔH_{lat}**.

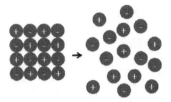

2. Interaction between the ions and the solvent molecules (to form **hydration shells**) which involves the **enthalpy of hydration, ΔH_{hyd}**.

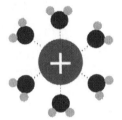 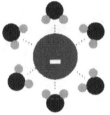

Positive cations attract the δ- oxygen atoms in the water.   Negative anions attract the δ+ hydrogen atoms in the water.

Lattice enthalpy have positive values (in line with the Data Booklet). Enthalpies of hydration are always exothermic and have negative values.
Enthalpy of solution, ΔH_{sol} can be calculated using the following relationship and dissolution energy cycles:
$\Delta H_{sol} = \Delta H_{lat} + \Delta H_{hyd}(\text{cation}) + \Delta H_{hyd}(\text{anion})$

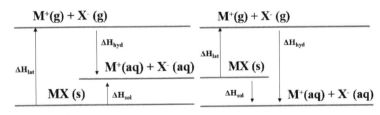

Endothermic dissolving   Exothermic dissolving
$|\Delta H_{lat}| > |\Delta H_{hyd}|$ $|\Delta H_{lat}| < |\Delta H_{hyd}|$

Solvation can be an endothermic or exothermic process and this depends on the relative magnitudes of the lattice enthalpy and enthalpy of hydration.

Factors affecting the enthalpy of hydration

Enthalpy of hydration, ΔH_{hyd} increases with:
- smaller ionic radius.
- larger ionic charge.

In both cases, the increased charge density on the ion leads to a higher degree of ordering of the solvent molecules.

Example: Calculate the enthalpy change of solution for magnesium chloride from the following data.

Enthalpy change	Value / kJ mol^{-1}
ΔH_{lat} (MgCl$_2$)	+2540
ΔH_{hyd} (Mg^{2+})	-1963
ΔH_{hyd} (Cl$^-$)	-359

$\Delta H_{sol}(MgCl_2) = \Delta H_{lat}(MgCl_2) + \Delta H_{hyd}(Mg^{2+}) + 2\Delta H_{hyd}(Cl^-)$
$\Delta H_{sol}(MgCl_2) = +2540 - 1963 - (2 \times 359) = -141$ kJ mol^{-1}

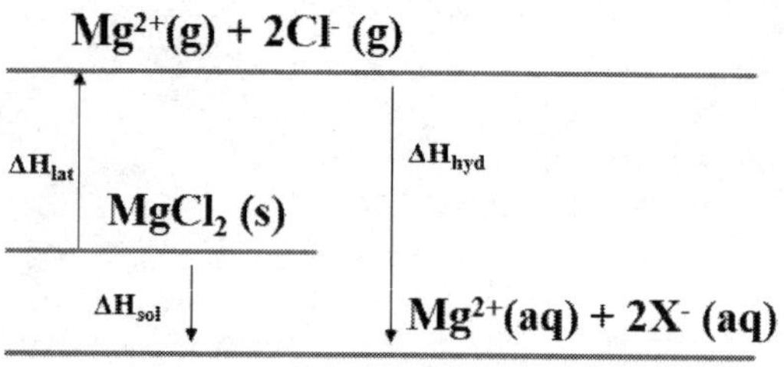

$Mg^{2+}(g) + 2Cl^-(g)$

ΔH_{lat} ΔH_{hyd}

MgCl$_2$ (s)

ΔH_{sol} $Mg^{2+}(aq) + 2X^-(aq)$

15.2 Entropy and spontaneity
Syllabus
Nature of science – can you relate this topic to these concepts?
Theories can be superseded—the idea of entropy has evolved through the years as a result of developments in statistics and probability.
Understandings – how well can you explain these statements?
Entropy (S) refers to the distribution of available energy among the particles. The more ways the energy can be distributed the higher the entropy.
Gibbs free energy (G) relates the energy that can be obtained from a chemical reaction to the change in enthalpy (ΔH), change in entropy (ΔS), and absolute temperature (T). Entropy of gas>liquid>solid under same conditions.
Applications and skills – how well can you do all of the following?
Prediction of whether a change will result in an increase or decrease in entropy by considering the states of the reactants and products.
Calculation of entropy changes (ΔS) from given standard entropy values ($S°$). Application of $\Delta G° = \Delta H° - T\Delta S°$ in predicting spontaneity and calculation of various conditions of enthalpy and temperature that will affect this. Relation of ΔG to position of equilibrium.

The Laws of Thermodynamics
First Law: Energy cannot be created or destroyed – it can only be converted into different forms (Law of Conservation of Energy).
Second Law: In any spontaneous process, the total entropy change is positive ($\Delta S_{total} > 0$). There is an increase in entropy or disorder.
Third Law: The entropy of any perfectly crystalline substance at a temperature of absolute zero (0 K or -273.16°C) is zero.
Entropy (S) is the measure of randomness or disorder and the total energy available between the particles. Disorder is considered to be greater when the energy is widespread amongst the particles. The Second Law of Thermodynamics determines:
- if a physical/chemical change is likely to happen at a particular temperature.
- if a redox reaction will take place.
- the position of equilibrium.

Entropy change of a reaction (ΔS)

Entropy (S) is state function (relative to zero entropy – defined in the Third Law of Thermodynamics). It is always positive.

The entropy of a substance depends on:
- temperature – entropy increases with temperature.
- phase – entropy increases when the molecules are more randomly arranged.
    - S (gas) > S (liquid) > S (solid)
- complexity – entropy increases when the molecule is more complex e.g.

S [$CaCO_3$ (s)] > S [CaO (s)]
S [NH_3 (g)] > S [H_2 (g)]

ΔS (reaction) = ΣS (products) − ΣS (reactants)

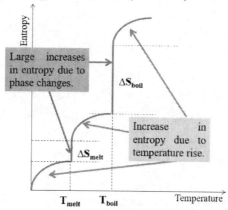

Predict whether the following will have a positive or negative entropy change (ΔS).

Combustion of phosphorus
P_4 (s) + 5O_2 (g) → P_4O_{10} (s)
6 molecules → 1 molecule
1 solid + 5 gas → 1 solid
Disorder decreases, ΔS < 0.

Calcium carbonate and hydrochloric acid
$CaCO_3$ (s) + 2HCl (aq) → $CaCl_2$ (aq) + H_2O (l) + CO_2 (g)
3 molecules → 3 molecules
1 solid + 2 liquid → 1 solid + 1 liquid + 1 gas
Disorder increases, ΔS > 0.

Compound	S° /J K⁻¹ mol⁻¹	Compound	S° /J K⁻¹ mol⁻¹
Solids		Gases	
C (graphite)	5.7	H_2	130.7
Mg	32.7	HCl	186.9
MgO	26.9	O_2	205.1
NaCl	72.1	Cl_2	223.1
I_2	116.1	Br_2	245.5
Liquids		CO	197.7
Hg	76.0	H_2O	188.8
Br_2	152.2	CO_2	213.8
H_2O	70.0	CH_4	186
C_2H_5OH	161	NH_3	192.5

ΔS (reaction) = ΣS (products) − ΣS (reactants)
Example:
$2Mg\ (s) + O_2\ (g) \rightarrow 2MgO\ (s)$
$\Delta S = 2 \times 26.9 - [2 \times 32.7 + 205.1]$
$\Delta S = -216.7$ J K⁻¹ mol⁻¹

Calculate ΔS for the following reactions:
$H_2\ (g) + Cl_2\ (g) \rightarrow 2HCl\ (g)$
$2H_2\ (g) + O_2\ (g) \rightarrow 2H_2O\ (l)$
$CH_4\ (g) + 2O_2\ (g) \rightarrow CO_2\ (g) + 2H_2O\ (l)$
Write the chemical equation for the combustion of ethanol and calculate ΔS.

Gibbs free energy

In order for a change to be spontaneous at constant temperature and pressure, the Gibbs free energy must be negative ($\Delta G < 0$). The Gibbs free energy is related to the enthalpy and entropy changes by the following equation:

$$\Delta G° = \Delta H° - T\Delta S°$$

As the temperature, T is the absolute temperature (in K), the value of T is always positive (assuming that absolute zero is not reachable). Therefore, the spontaneity of a reaction at specific temperatures can be predicted by these relationships.

ΔH	ΔS	ΔG	Spontaneity
> 0	> 0	> 0 at low T	not spontaneous as $\Delta H > T\Delta S$
		< 0 at high T	spontaneous as $\Delta H < T\Delta S$
> 0	< 0	> 0	forward reaction not spontaneous at all temperatures
			reverse reaction spontaneous at all temperatures
< 0	> 0	< 0	forward reaction spontaneous at all temperatures
			reverse reaction spontaneous at all temperatures
< 0	< 0	> 0 at high T	not spontaneous as $T\Delta S > \Delta H$
		< 0 at low T	spontaneous as $T\Delta S > \Delta H$

Gibbs free energy – further explanation ($\Delta H > 0$ and $\Delta S > 0$)

$CaCO_3\ (s) \rightarrow CaO\ (s) + CO_2\ (g) \quad \Delta H = +178\ kJ\ mol^{-1}$

There is an increase in entropy because of the formation of the gas.

$\Delta G° = \Delta H° - T\Delta S°$

$\Delta H > 0$;

$\Delta S > 0$ therefore $-T\Delta S < 0$

At low temperatures, the $-T\Delta S$ probably will not be negative enough to compensate for the positive ΔH, and so ΔG will be positive. The reaction is not spontaneous.

As T increases, the $-T\Delta S$ becomes more negative, and will eventually cancel out the positive ΔH. At that point the reaction will become spontaneous.

The reaction is not spontaneous at low temperatures, but will become spontaneous at higher ones.

Gibbs free energy – further explanation ($\Delta H > 0$ and $\Delta S < 0$)

$6C\ (s) + 3H_2\ (g) \rightarrow C_6H_6\ (l) \quad \Delta H = +49\ kJ\ mol^{-1}$

The entropy change will be negative because 3 moles of gas react to form 1 mole of liquid. The carbon is so highly ordered that its entropy is negligible.
$\Delta G° = \Delta H° - T\Delta S°$
$\Delta H > 0$;
$\Delta S < 0$ therefore $-T\Delta S > 0$
$-T\Delta S$ becomes even more positive as T increases.
ΔG is always positive at all temperatures (because two positive numbers are being added together).
The reaction is never spontaneous at any temperature.

Gibbs free energy – further explanation ($\Delta H < 0$ and $\Delta S > 0$)

$2C\ (s) + O_2\ (g) \rightarrow 2CO\ (g)$ $\Delta H = -221\ kJ\ mol^{-1}$
The entropy change will be positive because more gas is generated (1 mole to 2 moles). The carbon is so highly ordered that its entropy is negligible.
$\Delta G° = \Delta H° - T\Delta S°$
$\Delta H < 0$;
$\Delta S > 0$ therefore $-T\Delta S < 0$
$-T\Delta S$ becomes even more negative as T increases.
ΔG is always negative at all temperatures (because two negative numbers are being added together).
The reaction is always spontaneous at any temperature but becomes even more spontaneous at higher temperatures.

Gibbs free energy – further explanation ($\Delta H < 0$ and $\Delta S < 0$)

$2Mg\ (s) + O_2\ (g) \rightarrow 2MgO\ (s)$ $\Delta H = -1203\ kJ\ mol^{-1}$
The entropy has decreased because of the loss of a mole of gas.
$\Delta G° = \Delta H° - T\Delta S°$
$\Delta H < 0$;
$\Delta S < 0$ therefore $-T\Delta S > 0$
$-T\Delta S$ becomes more positive as T increases.
Because of the very high negative value of ΔH, ΔG will be negative at ordinary temperatures, and so the reaction will be spontaneous.
At very high temperatures, the positive value of $-T\Delta S$ will eventually cancel out the negative value of ΔH.
ΔG will then become positive, and so the reaction will cease to be spontaneous.

Gibbs free energy change (ΔG) calculations

Example 1:
Show that the thermal decomposition of sodium carbonate is not feasible at 1200 K. At what temperature does it become spontaneous?
$Na_2CO_3 (s) \rightarrow Na_2O (s) + CO_2(g)$
$\Delta H = +323.1$ kJ mol^{-1}
$\Delta S = +153.7$ J K^{-1} mol^{-1}
Convert the ΔS units from joules to kilojoules by dividing by 1000. In this case, then, $\Delta S = +0.1537$ kJ K^{-1} mol^{-1}
$\Delta G = \Delta H - T\Delta S = +323.1 - (1200 \times +0.1537) = +138.7$ kJ mol^{-1}
Because $\Delta G > 0$ at 1200 K, the reaction is not spontaneous.
To find the temperature at which the reaction becomes spontaneous,
$\Delta H - T\Delta S < 0$
$323.1 - 0.1537T < 0$
$0.1537T > 323.1$
$T > 2102$ K
Therefore, the reaction will not be spontaneous until it reaches a temperature over 2102 K.

Example 2:
Two reactions involved in the extraction of iron in the blast furnace are:
A: $Fe_2O_3 (s) + 3CO (g) \rightarrow 2Fe (s) + 3CO_2 (g)$
$\Delta H = -25.0$ kJ mol^{-1}
$\Delta S = +15.2$ J K^{-1} mol^{-1}
B: $Fe_2O_3 (s) + 3C (g) \rightarrow 2Fe (s) + 3CO (g)$
$\Delta H = +491$ kJ mol^{-1}
$\Delta S = +542.9$ J K^{-1} mol^{-1}
Which reaction is more spontaneous at (a) 800 K, (b) 1800 K?
(a) at 800 K
    A: $\Delta G = -25.0 - (800 \times 0.0152) = -37.2$ kJ mol^{-1}
    B: $\Delta G = 491 - (800 \times 0.5429) = +56.7$ kJ mol^{-1}
       A is spontaneous but B is not.
(b) at 1800 K
    A: $\Delta G = -25.0 - (1800 \times 0.0152) = -52.4$ kJ mol^{-1}
    B: $\Delta G = 491 - (1800 \times 0.5429) = -486$ kJ mol^{-1}
       Both reactions are spontaneous at 1800 K.

Gibbs free energy change of formation (ΔG_f)

This is the free energy change when 1 mol of a substance is formed from its elements which are in their normal states under standard conditions of 298 K and 1.00×10^5 Pa. The values which are given in the Data Booklet can be used to calculate the Gibbs free energy change for a reaction using the following relationship:
$\Delta G_r = \Sigma \Delta G_f$ **(products)** - $\Sigma \Delta G_f$ **(reactants)**

Example 1:
What is the Gibbs free energy change for the manufacture of ethanol from ethene and steam?
C_2H_4 (g) + H_2O (g) → C_2H_5OH (l) ΔH = -45 kJ mol^{-1}
ΔG_f (C_2H_4 (g)) = -68.0 kJ mol^{-1}
ΔG_f (H_2O (g)) = -228.6 kJ mol^{-1}
ΔG_f (C_2H_5OH (l)) = -175 kJ mol^{-1}
ΔG_r = -175 – [(-68.0 + (-228.6)] = **+121.6 kJ mol^{-1}**
It can be seen that ΔS < 0 as two gaseous molecules form one liquid product.
As ΔH < 0, ΔG will have a negative value at lower temperatures. Therefore, the forward reaction is more spontaneous at low temperatures.

Example 2:
What is the Gibbs free energy change for the combustion of propane?
C_3H_8 (g) + $5O_2$ (g) → $3CO_2$ (g) + $4H_2O$ (l)
ΔH = -2219 kJ mol^{-1}
ΔG_f (C_3H_8 (g)) = -24.0 kJ mol^{-1}
ΔG_f (O_2 (g)) = 0 kJ mol^{-1}
ΔG_f (CO_2 (g)) = -394.4 kJ mol^{-1}
ΔG_f (H_2O (l)) = -237.1 kJ mol^{-1}
ΔG_r = [(3×-394.4) + (4×-237.1)] – (-24.0) = **-2107.6 kJ mol^{-1}**
It can be seen that ΔS < 0 as there are fewer gaseous molecules on the right side.
As ΔH < 0, ΔG will have a negative value at lower temperatures. Therefore, the forward reaction is more spontaneous at low temperatures. However, the reaction does not happen on its own as it requires a high activation energy.

Gibbs free energy change and extent of reversible reactions

Consider the manufacture of ammonia which is an exothermic process:

$N_2 (g) + 3H_2 (g) \rightleftharpoons 2NH_3 (g)$

Forward reaction:
- Gibbs free energy decreases towards a minimum value which is reached at equilibrium.
- forward reaction is spontaneous and reverse reaction is non-spontaneous.

Reverse reaction:
- Gibbs free energy decreases towards a minimum value which is reached at equilibrium.
- reverse reaction is spontaneous and forward reaction is non-spontaneous.

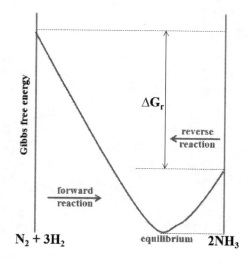

6.1 Collision theory and rates of reaction

Syllabus

Nature of science – can you relate this topic to these concepts?
The principle of Occam's razor is used as a guide to developing a theory – although we cannot directly see reactions taking place at the molecular level, we can theorize based on the current atomic models. Collision theory is a good example of this principle.

Understandings – how well can you explain these statements?
Species react as a result of collisions of sufficient energy and proper orientation.

The rate of reaction is expressed as the change in concentration of a particular reactant/product per unit time.

Concentration changes in a reaction can be followed indirectly by monitoring changes in mass, volume and colour.

Activation energy (E_a) is the minimum energy that colliding molecules need in order to have successful collisions leading to a reaction.

By decreasing E_a, a catalyst increases the rate of a chemical reaction, without itself being permanently chemically changed.

Applications and skills – how well can you do all of the following?
Description of the kinetic theory in terms of the movement of particles whose average kinetic energy is proportional to temperature in Kelvin.

Analysis of graphical and numerical data from rate experiments.

Explanation of the effects of temperature, pressure/concentration and particle size on rate of reaction.

Construction of Maxwell–Boltzmann energy distribution curves to account for the probability of successful collisions and factors affecting these, including the effect of a catalyst.

Investigation of rates of reaction experimentally and evaluation of the results.

Sketching and explanation of energy profiles with and without catalysts.

Rates of reaction

The rate of a chemical reaction is the change in concentration of the reactants or products over a period of time.

The units for concentration (c) are always mol dm^{-3}. Therefore the units for rate of reaction is mol dm^{-3} s^{-1}.

$$\text{rate of reaction} = \frac{\Delta c}{\Delta t}$$

Change of concentration of reactants
The concentration of the reactant(s) decrease over time – the rate of reaction has a negative value.

Change of concentration of products
The concentration of the product(s) increase over time – the rate of reaction has a positive value.

Concentration time graph

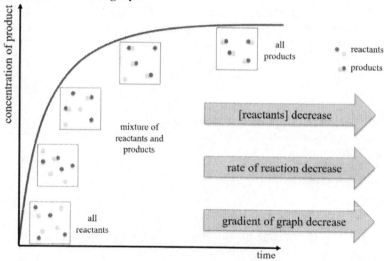

Tracking [product]

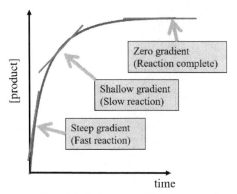

Tracking [reactant]

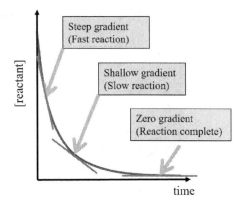

Concentration time graphs (typical examples)

A → B

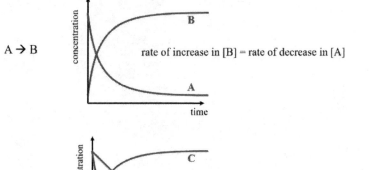

rate of increase in [B] = rate of decrease in [A]

A + 2B → C

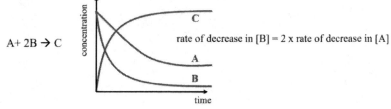

rate of decrease in [B] = 2 x rate of decrease in [A]

Measuring rates of reaction

Method 1: Average rate
This is the ratio of the change in concentration of reactant or product to a given time interval (usually until completion of the reaction).

$$\text{average rate} = \frac{\Delta c}{\Delta t}$$

Method 2: Instantaneous rate
This is the ratio of the change in concentration of reactant or product to a time interval at a specific point during the reaction.

$$\lim_{\Delta t \to 0} \frac{\Delta c}{\Delta t} = \frac{dc}{dt}$$

Method 3: Initial rate
This is the ratio of the change in concentration of reactant or product to a time interval at the start of the reaction (t=0).
The gradient of the tangent to the curve at a particular time gives the rate of reaction at that instant, therefore the gradient of the tangent to the curve at 0s gives the initial rate.

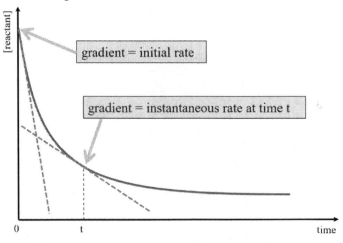

Measuring rates of reaction – Gas evolved

The volume of gas evolved is proportional to the number of moles of gas produced. The gas can be collected by:
- gas syringe

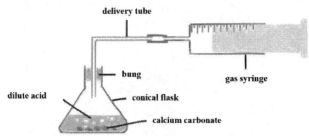

- water displacement in an inverted burette. This method is convenient and effective when the gas has low solubility in water e.g. hydrogen. With gases like carbon dioxide, solubility can be decreased by using warm water.

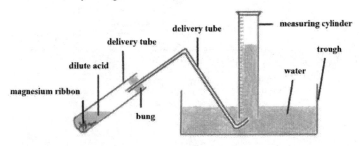

Example:
Mg (s) + 2HCl (aq) → MgCl$_2$ (aq) + H$_2$ (g)
CaCO$_3$ (s) + 2HCl (aq) → CaCl$_2$ (aq) + H$_2$O (l) + CO$_2$ (g)

t (±0.01s)	0.00	20.00	40.00	60.00	80.00	100.00
V (±0.5 cm^3)	0.0	71.0	106.0	132.0	155.0	168.0

t (±0.01s)	100.00	120.00	140.00	160.00	180.00	200.00
V (±0.5 cm^3)	168.0	176.0	179.0	179.0	179.0	179.0

Average rate
Average rate = 179.0 / 140.00 = 1.28 cm^3 s^{-1}
Initial rate
Initial rate = 130.0 / 36.00 = 3.61 cm^3 s^{-1}
Instantaneous rate at 102.00s
Rate = 52.0 / 72.00 = 0.72 cm^3 s^{-1}

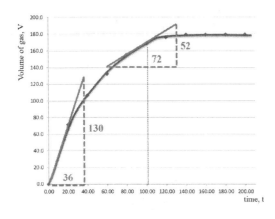

Measuring rates – Loss of mass

The reactions above can also be modified to measure the loss of mass as gas is released into the air. This method is not suitable for very light gases such as hydrogen as the change in mass will be too insignificant.

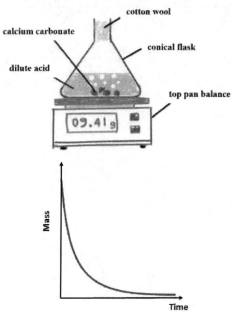

Measuring rates of reaction – conductivity

The presence of ions allows a solution to conduct electricity, so if there is a significant change in the concentration of ions (especially hydrogen and hydroxide ions which have an unusually high conductivity) during the course of a reaction, then the reaction rate may be found from the change in conductivity. This can be tracked using a conductivity probe and meter (the A.C. resistance between two inert electrodes immersed in the solution is measured). Examples:

PCl_3 (aq) + $3H_2O$ (l) → $H_2PO_3^-$ (aq) + $4H^+$ (aq) + $3Cl^-$ (aq)
The electrical conductivity increases as the concentration of ions increases from 0 to 8.

BrO_3^- (aq) + $5Br^-$ (aq) + $6H^+$ (aq) → $3Br_2$ (aq) + $3H_2O$ (l)
The electrical conductivity decreases as the concentration of ions decreases from 12 to 0.

Measuring rates – Colorimetry

The amount of light of a particular **frequency** that is **absorbed** depends on the **concentration** of the **coloured substance**. The coloured substance is either the reactant, product or an indicator. The light transmitted is measured at regular intervals and recorded on a computer or datalogger.

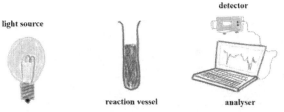

Example:
Br_2 (aq) + HCOOH (aq) → $2Br^-$ (aq) + CO_2 (g) + $2H^+$ (aq)
The colour of bromine fades gradually.
2HI (g) → H_2 (g) + I_2 (g)
The colour of iodine increases in intensity.

Measuring rates of reaction – change in pH

Acid-base reactions can be followed by measuring the pH of the reaction mixture over time. The pH of the solution is a measure of the concentration of hydronium ions, H_3O^+ which is measured using a pH probe and meter. Examples:
HNO_3 (aq) + NaOH (aq) → $NaNO_3$ (aq) + H_2O (l)
H_2SO_4 (aq) + $2NaHCO_3$ (s) → Na_2SO_4 (aq) + $2H_2O$ (l) + $2CO_2$ (g)

Collision Theory

In order for a chemical reaction to happen:
- Particles **must collide** with each other.
  Only a certain fraction of the total collisions cause chemical change; these are called **successful collisions**. In order for collisions to be successful, they must have:
    - sufficient **kinetic energy** to initiate the reaction. The minimum energy required by colliding particles for a successful reaction is the **activation energy (E_a)**.
    - the **correct orientations with respect to each other** – a head-on collision has a greater chance of resulting in a reaction.

The majority of collisions do not lead to a reaction.

Criteria 1 – Energy of collision
Exothermic reaction

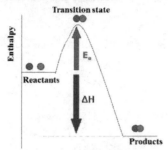

Endothermic reaction

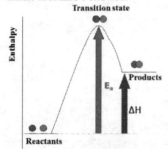

The reaction profile diagrams show the energy levels of the reactants, transition state and products of the reaction. The enthalpy change, ΔH and the activation energy, E_a of the reaction are shown on the diagram.

Maxwell-Boltzmann distribution

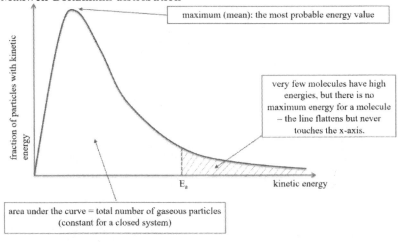

Criteria 2 – Geometry of collision
Glancing blow collision
Reactants collide but no reaction occurs

Head-on collision
Reactants collide, reaction occurs and product(s) are formed

Factors affecting reaction rates
Rates of reactions can be affected by the following factors:
- Pressure
- Concentration
- Particle size (surface area)
- Temperature
- Catalyst

Pressure
This applies to reactions involving **gases**.
The pressure in a system can be increased by:
- **decreasing the volume** of the container.
- pumping **more reactant gas** into the reacting vessel.

Increasing the pressure in a system increases the rate of reaction because:
- the number of particles per volume increases.
- the **frequency of collision** between the particles **increases**.

The average kinetic energy remains the same, so the proportion of collisions which result in reaction is the same.

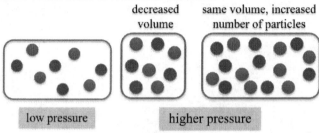

Concentration

This applies to reactions involving **gases** and **solutions**.
The concentration of a reactant can be increased by:
- **increasing the amount of solute in a given volume**.

Increasing the concentration of a reactant increases the rate of reaction because:
- the number of colliding particles per volume increases.
- the **frequency of collision** between the particles **increases**.

The average kinetic energy remains the same, so the proportion of collisions which result in reaction is the same.

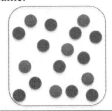

There are fewer red particles inc the same volume so there is less chance of a collision.

There are more red particles in the same volume so there is more chance of a collision so the reaction goes faster.

Particle size (surface area)
This applies to **heterogeneous** systems involving at least one **solid**. Only the particles which are on the surface of the solid are in contact with the liquid or gas.
The **surface area** of a reactant can be **increased** by:
- **decreasing** the **particle size** (smaller pieces or powdered).

Increasing the surface area of a reactant increases the rate of reaction because:
- the number of particles exposed for collision increases.
- the **frequency of collision** between the particles **increases**.

large particle

number of particles exposed for collision
= 6 × 4 × 4 = 96

same amount cut into 8 smaller pieces

8 ×

number of particles exposed for collision
= 8 × 6 × 2 × 2 = 192

Temperature
This applies to **all reactions**.
Increasing the temperature in a system increases the rate of reaction because:
- the **average kinetic energy** of the particles is **proportional to the temperature of the system in kelvin** – a greater fraction of particles have the energy necessary (higher than the activation energy) to react on collision. The shaded area in the Maxwell-Boltzmann energy distribution curve increases.
- the frequency of collision between the particles is increases (minor factor).

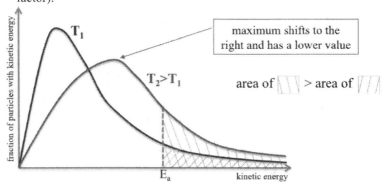

Catalyst
Catalysts increases the rate of reaction by:
- providing an **alternative path** for the reaction and **lowers the activation energy.**
- the catalyst is **not consumed** in the reaction itself.

Many biochemical processes are aided by biological catalysts called **enzymes**.

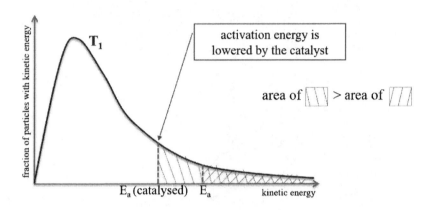

Exothermic reactions

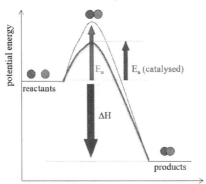

Endothermic reactions

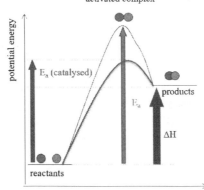

Homogeneous catalysts
Catalysts which are in the same phase as the reactants.

Example 1: Oxidation of iodide ions, I^- by persulfate ions, $S_2O_8^{2-}$ catalysed by iron (II) ions, Fe^{2+}.
Overall reaction:
$S_2O_8^{2-}$ (aq) + $2I^-$ (aq) → $2SO_4^{2-}$ (aq) + I_2 (s)
This reaction is slow without a catalyst because it requires the collision of three negative ions which repel one another.
1st step – formation of intermediate compound (Fe^{3+}):
$S_2O_8^{2-}$ (aq) + $2Fe^{2+}$ (aq) → $2SO_4^{2-}$ (aq) + $2Fe^{3+}$ (aq)
2nd step – intermediate reacts with the other reactant:
$2Fe^{3+}$ (aq) + $2I^-$ (aq) → I_2 (s) + $2Fe^{2+}$ (aq)

Example 2: destruction of atmospheric ozone (O_3) by chlorine radicals (Cl•).
Natural ozone destruction and production (conversion of UV radiation to heat):
1. Destruction by solar UV radiation: O_3 (g) → O_2 (g) + O• (g)
2. Oxygen free radical attacks oxygen molecules: O_2 (g) + O• (g) → O_3 (g) + heat

High frequency UV-a and UV-b radiation has been linked with increased cases of melanoma (skin cancer) and cataracts. The presence of ozone in the stratosphere absorbs 95% of this harmful radiation. Chlorofluorocarbons (CFCs) such as freon (CF_2Cl_2) are inert in the troposphere but higher energy UV radiation in the stratosphere causes the C-Cl bonds to break homolytically. The C-F bonds are stronger due to its higher polarity. Chlorine free radicals (Cl•) are released which attack ozone molecules in a chain reaction.
CF_2Cl_2 (g) → CF_2Cl• (g) + Cl• (g)
Overall reaction: $2O_3$ (g) → $3O_2$ (g)

Free radical mechanism between chlorine and ozone:
Step 1: Cl• (g) + O_3 (g) → ClO• (g) + O_2 (g)
Step 2: ClO• (g) + O_3 (g) → $2O_2$ (g) + Cl• (g)
The chlorine free radical (Cl•) is regenerated in the second step i.e. each chlorine free radical has the potential to destroy two ozone molecules.
Although ozone is essential as **a UV filter** in the stratosphere, its presence in the troposphere can lead to **respiratory problems** e.g. asthma and emphysema as well as contribute to the **greenhouse effect**. **Nitrogen oxides (NO_x)** from car exhausts can react with **volatile organic compounds (VOCs)** to form ozone.

Heterogeneous catalysts
Catalysts which are in a different phase from the reactants.

Example 1: Haber process catalysed by iron
$$N_2(g) + 3H_2(g) \xrightarrow{Fe(s)} 2NH_3(g)$$

Example 2: Contact process catalysed by vanadium(V) oxide
$$2SO_2(g) + O_2 \xrightarrow{V_2O_5(s)} 2SO_3(g)$$
The reactants are **adsorbed** on the **active sites** of the catalyst where reaction takes place.

Example 3: catalytic converter in car exhaust
Typical catalysts are **platinum, palladium, rhodium, V_2O_5, CuO and Cr_2O_3**. Cars fitted with catalytic converters must run on **unleaded fuel**. The main pollutants in a car engine are:
- **nitrogen oxides (NO and NO_2) from** the reaction on atmospheric nitrogen (N_2) and oxygen (O_2) at **high temperature** in the engine – these gases contribute to the formation **of acid rain**.
- **carbon monoxide (CO)** from the incomplete combustion of fuels – this gas is **toxic** and leads to **blood poisoning**.
- **unburned hydrocarbons (C_xH_y)** due to insufficient supply of oxygen in the engine – these are **greenhouse gases**.

These are converted into less harmful N_2, CO_2 and H_2O by the catalytic converter:
$$2NO(g) \rightarrow N_2(g) + O_2(g)$$
$$2NO_2(g) \rightarrow N_2(g) + 2O_2(g)$$
$$2CO(g) + O_2(g) \rightarrow 2CO_2(g)$$
$$C_3H_8(g) + 5O_2(g) \rightarrow 3CO_2(g) + 4H_2O(g)$$
$$2C_4H_{10}(g) + 13O_2(g) \rightarrow 8CO_2(g) + 10H_2O(g)$$

16.1 Rate expression and reaction mechanism
Syllabus
Nature of science – can you relate this topic to these concepts?
Principle of Occam's razor – newer theories need to remain as simple as possible while maximizing explanatory power. The low probability of three molecule collisions means stepwise reaction mechanisms are more likely.

Understandings – how well can you explain these statements?
Reactions may occur by more than one step and the slowest step determines the rate of reaction (rate determining step/RDS).

The molecularity of an elementary step is the number of reactant particles taking part in that step.

The order of a reaction can be either integer or fractional in nature. The order of a reaction can describe, with respect to a reactant, the number of particles taking part in the rate-determining step.

Rate equations can only be determined experimentally.

The value of the rate constant (k) is affected by temperature and its units are determined from the overall order of the reaction.

Catalysts alter a reaction mechanism, introducing a step with lower activation energy.

Applications and skills – how well can you do all of the following?
Deduction of the rate expression for an equation from experimental data and solving problems involving the rate expression.

Sketching, identifying, and analysing graphical representations for zero, first and second order reactions.

Evaluation of proposed reaction mechanisms to be consistent with kinetic and stoichiometric data.

Rate equation
Consider a reaction where the rate of reaction depends on the concentrations of the reactants A and B.

The rate equation is given as: rate = $k[A]^x[B]^y$

[A] = concentration of reactant A
[B] = concentration of reactant B
k = rate constant
x and y = partial orders of reaction with respect to A and B
x+y = overall order of reaction (the sum of the exponents to which the concentrations of the reactants are raised in the rate equation)

Rate equations must be derived experimentally. They cannot be determined by looking at the coefficients in the balanced chemical equation!
The most common method for determining the rate of reaction is by using the **initial rates** method.

Example 1:
In a reaction involving reagents A and B, the concentrations of A and B were varied independently. The corresponding initial rates of the reaction were recorded as such:

Experiment	[A] / mol dm^{-3}	[B] / mol dm^{-3}	Initial rate / mol dm^{-3} s^{-1}
1	0.50	1.00	1.00
2	1.00	1.00	2.00
3	1.00	2.00	2.00

Experiment 1 and 2: [A] was doubled, [B] remained constant.
The initial rate of reaction doubled. This means that the rate of reaction is proportional to [A] or rate α $[A]^1$. The order with respect to A is 1.
Experiment 2 and 3: [A] remained constant, [B] was doubled.
The initial rate of reaction stayed the same. This means that the rate of reaction does not depend on [B] or rate α $[B]^0$. The order with respect to B is 0.
rate = $k[A]^1[B]^0$ or rate = k[A]
Overall order of reaction = 1 + 0 = 1

Example 2:

Experiment	[A] / mol dm^{-3}	[B] / mol dm^{-3}	Initial rate / mol dm^{-3} s^{-1}
1	1.10×10^{-2}	6.25×10^{-3}	1.00
2	1.10×10^{-2}	1.25×10^{-2}	2.00
3	2.20×10^{-2}	6.25×10^{-3}	2.00

Experiment 1 and 2: [A] remained constant, [B] was doubled.
The initial rate of reaction doubled. This means that the rate of reaction is proportional to [B] or rate α [B]^1. The order with respect to B is 1.
Experiment 1 and 3: [A] was doubled, [B] remained constant.
The initial rate of reaction doubled. This means that the rate of reaction is proportional to [A] or rate α [A]^1. The order with respect to A is 1.
rate = k[A]^1[B]^1 or rate = k[A][B]
Overall order of reaction = 1 + 1 = 2

Example 3:

Experiment	[A] / mol dm^{-3}	[B] / mol dm^{-3}	Initial rate / mol dm^{-3} s^{-1}
1	3.25×10^{-4}	8.92×10^{-4}	5.82×10^{-6}
2	3.25×10^{-4}	1.78×10^{-3}	1.16×10^{-5}
3	6.50×10^{-4}	8.92×10^{-4}	2.32×10^{-5}

Experiment 1 and 2: [A] remained constant, [B] was doubled.
The initial rate of reaction doubled. This means that the rate of reaction is proportional to [B] or rate α [B]^1. The order with respect to B is 1.
Experiment 1 and 3: [A] was doubled, [B] remained constant.
The initial rate of reaction quadrupled. This means that the rate of reaction is proportional to [A]^2 or rate α [A]^2. The order with respect to A is 2.
rate = k[A]^2[B]^1 or rate = k[A]^2[B]
Overall order of reaction = 2 + 1 = 3

Example 4:

[X] / mol dm^{-3}	[Y] / mol dm^{-3}	[Z] / mol dm^{-3}	Initial rate / mol dm^{-3} s^{-1}
0.10	0.10	0.10	2.40 × 10^{-3}
0.10	0.10	0.30	7.20 × 10^{-3}
0.05	0.10	0.10	2.40 × 10^{-3}
0.10	0.40	0.10	3.84 × 10^{-2}

[X] and [Y] remained constant, [Z] was tripled.
The initial rate of reaction tripled. This means that the rate of reaction is proportional to [Z] or rate α [Z]^1. The order with respect to Z is 1.
[X] was halved, [Y] and [Z] remained constant.
The initial rate of reaction stayed the same. This means that the rate of reaction does not depend on [X] or rate α [X]^0. The order with respect to X is 0.
[X] and [Z] remained constant, [Y] was quadrupled.
The initial rate of reaction was 16 times faster. This means that the rate of reaction is proportional to [Y]^2 or rate α [Y]^2. The order with respect to Y is 2.
rate = k[X]^0[Y]^2[Z]^1 or rate = k[Y]^2[Z]
Overall order of reaction = 0 + 2 + 1 = 3

Units of rate constants
As the rate of reaction is defined as the change in concentration over time, the units for rate are always **mol dm^{-3} s^{-1}**.
The units for concentration are always **mol dm^{-3}**.
Therefore, the units for the rate constant, k varies according to the orders of reaction with respect to the corresponding reactants.

Zero-order reactions
rate = k [A]^0
rate = k
Unit for k is **mol dm^{-3} s^{-1}**

First order reactions
rate = k [A]^1
rate / [A] = k
Unit for k = $\dfrac{\text{mol dm}^{-3}\text{ s}^{-1}}{\text{mol dm}^{-3}}$ = **s^{-1}**

Second order reactions
rate = k [A]^2
rate / [A]^2 = k
Unit for k = $\dfrac{\text{mol dm}^{-3}\text{ s}^{-1}}{\text{mol}^2\text{ dm}^{-6}}$ = **mol^{-1} dm^3 s^{-1}**

The units for higher order reactions can be worked out in a similar way.

Concentration-time graphs for zero, first and second order reactions

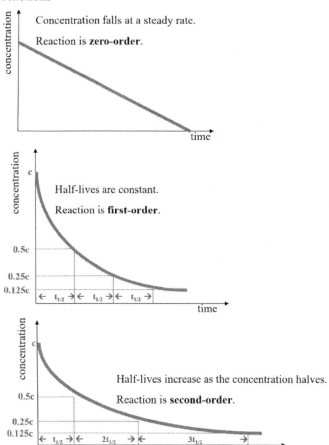

Rate-concentration graphs for zero, first and second order reactions

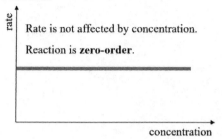

Rate is not affected by concentration.

Reaction is **zero-order**.

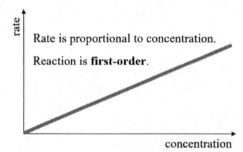

Rate is proportional to concentration.

Reaction is **first-order**.

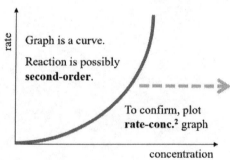

Graph is a curve.

Reaction is possibly **second-order**.

To confirm, plot **rate-conc.^{2}** graph

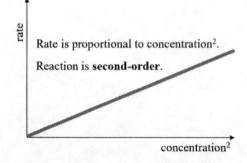

Rate is proportional to concentration^2.

Reaction is **second-order**.

Reaction mechanisms and molecularity
Consider the following reaction:
BrO_3^- (aq) + $6H^+$ (aq) + $5Br^-$ (aq) → $3Br_2$ (aq) + $3H_2O$ (l)
In order for this reaction to occur in one step, there must be a simultaneous collision between 12 ions, all at the appropriate mutual geometry and with sufficient kinetic energy – a highly unlikely event! Most reactions are multi-step. Each step (called an elementary step) has a rate and its own rate constant – these could be fast or slow.
Elementary steps can be:
- **unimolecular**: involving a single molecule.
- **bimolecular**: involving two molecules, atoms or ions.
- **termolecular**: involving three molecules, atoms or ions.

The slowest step of the reaction is called the **rate-determining step (rds)** and this step is slow because it has the highest **activation energy (E_a)**.
Fast steps before the rds affect the overall rate of reaction.
Fast steps after the rds do not affect the overall rate of reaction.

Consider the reaction with stoichiometry **2X + Y → Z**
Experimental data shows that it is **first order** with respect to both **X** and **Y** so the rate equation is **rate = k[X][Y]**
Possible mechanisms:
A
Step 1: X + Y → intermediate   (fast)
Step 2: intermediate + X → Z   (rds)
X appears twice and Y appears once up to and including the rds. This would suggest that it is second order with respect to X and first order with respect to Y. This is inconsistent with the data and is not the mechanism.
B
Step 1: X + Y → intermediate   (rds)
Step 2: intermediate + X → Z   (fast)
X and Y appear once up to and including the rds. This would suggest that it is first order with respect to X and Y. Step 2 occurs after the rds and so does not affect the overall rate of reaction. This is consistent with the data and is the likely mechanism.

Example 1:
Consider the reaction: $2NO_2\ (g) + F_2\ (g) \rightarrow 2NO_2F\ (g)$
Experimentally, this reaction is first order with respect to NO_2 and F_2. The rate equation is therefore rate = $k[NO_2][F_2]$

Suggestion 1: single step
$2NO_2\ (g) + F_2\ (g) \rightarrow 2NO_2F\ (g)$ termolecular rds
The rate equation for this mechanism would be rate = $k[NO_2]^2[F_2]$

Suggestion 2: two steps
Step 1: $NO_2\ (g) + F_2\ (g) \rightarrow NO_2F\ (g) + F$ bimolecular fast
Step 2: $NO_2\ (g) + F\ (g) \rightarrow NO_2F\ (g)$ bimolecular rds
The rate equation for this mechanism would be rate = $k[NO_2]^2[F_2]$

Suggestion 3: two steps
Step 1: $NO_2\ (g) + F_2\ (g) \rightarrow NO_2F\ (g) + F$ bimolecular rds
Step 2: $NO_2\ (g) + F\ (g) \rightarrow NO_2F\ (g)$ bimolecular fast
The rate equation for this mechanism would be rate = $k[NO_2][F_2]$
Suggestion 3 is therefore the correct mechanism.

Example 2:
Consider the reaction: $2NO\ (g) + O_2\ (g) \rightarrow 2NO_2\ (g)$
Experimentally, this reaction is first order with respect to O_2 and second order with respect to NO. The rate equation is therefore rate = $k[NO]^2[O_2]$

Suggestion 1: single step
$2NO\ (g) + O_2\ (g) \rightarrow 2NO_2\ (g)$ termolecular rds
The rate equation for this mechanism would be rate = $k[NO]^2[O_2]$

Suggestion 2: two steps
Step 1: $NO\ (g) + NO\ (g) \rightarrow N_2O_2\ (g)$ bimolecular fast
Step 2: $N_2O_2\ (g) + O_2\ (g) \rightarrow 2NO_2\ (g)$ bimolecular rds
The rate equation for this mechanism would be rate = $k[NO]^2[O_2]$

Suggestion 3: two steps
Step 1: $NO\ (g) + NO\ (g) \rightarrow N_2O_2\ (g)$ bimolecular rds
Step 2: $N_2O_2\ (g) + O_2\ (g) \rightarrow 2NO_2\ (g)$ bimolecular fast
The rate equation for this mechanism would be rate = $k[NO]^2$
Suggestion 2 is therefore the correct mechanism. Suggestion 1 is incorrect as termolecular elementary steps are almost impossible.

Example 3:
Consider the reaction: $H_2 (g) + 2ICl (g) \rightarrow I_2 (g) + 2HCl (g)$
Experimentally, this reaction is first order with respect to H_2 and ICl.
The rate equation is therefore rate = $k[H_2][ICl]$

Suggestion 1: single step
$H_2 (g) + 2ICl (g) \rightarrow I_2 (g) + 2HCl (g)$ termolecular rds
The rate equation for this mechanism would be rate = $k[H_2][ICl]^2$

Suggestion 2: two steps
Step 1: $H_2 (g) + ICl (g) \rightarrow HI (g) + HCl (g)$ bimolecular fast
Step 2: $HI (g) + ICl (g) \rightarrow I_2 (g) + HCl (g)$ bimolecular rds
The rate equation for this mechanism would be rate = $k[H_2][ICl]^2$

Suggestion 3: two steps
Step 1: $H_2 (g) + ICl (g) \rightarrow HI (g) + HCl (g)$ bimolecular rds
Step 2: $HI (g) + ICl (g) \rightarrow I_2 (g) + HCl (g)$ bimolecular slow
The rate equation for this mechanism would be rate = $k[H_2][ICl]$
Suggestion 3 is therefore the correct mechanism.

Example 4:
Consider the reaction: $2H_2O_2 (l) \rightarrow 2H_2O (l) + O_2 (g)$ (catalysed by I_2)
Experimentally, this reaction is first order with respect to H_2O_2 and I^-.
The rate equation is therefore rate = $k[H_2O_2][I^-]$

Suggestion 1: single step
$2H_2O_2 (l) \rightarrow 2H_2O (l) + O_2 (g)$
    bimolecular rds
The rate equation for this mechanism would be rate = $k[H_2O_2]^2$

Suggestion 2: two steps
Step 1: $H_2O_2 (l) + I^- (aq) \rightarrow H_2OH (aq) + IO^- (aq)$
    bimolecular rds
Step 2: $H_2O_2 (l) + IO^- (aq) \rightarrow H_2O (l) + O_2 (g) + I- (aq)$
    bimolecular fast
The rate equation for this mechanism would be rate = $k[H_2O_2][I^-]$

Suggestion 3: two steps
Step 1: $H_2O_2 (l) + I^- (aq) \rightarrow H_2OH (aq) + IO^- (aq)$
    bimolecular fast
Step 2: $H_2O_2 (l) + IO^- (aq) \rightarrow H_2O (l) + O_2 (g) + I- (aq)$
    bimolecular rds
The rate equation for this mechanism would be rate = $k[H_2O_2]^2[I^-]$
Suggestion 2 is therefore the correct mechanism.

Multi-step mechanisms

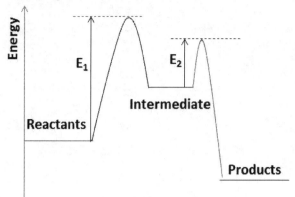

$E_1 > E_2$:
First step is rate determining.

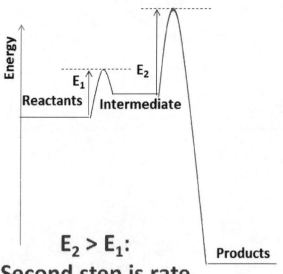

$E_2 > E_1$:
Second step is rate determining.

16.2 Activation energy
Syllabus
Nature of science – can you relate this topic to these concepts?
Theories can be supported or falsified and replaced by new theories – changing the temperature of a reaction has a much greater effect on the rate of reaction than can be explained by its effect on collision rates. This resulted in the development of the Arrhenius equation, which proposes a quantitative model to explain the effect of temperature change on reaction rate.

Understandings – how well can you explain these statements?
The Arrhenius equation uses the temperature dependence of the rate constant to determine the activation energy.
A graph of $1/T$ against $\ln k$ is a linear plot with gradient $- E_a / R$ and intercept, $\ln A$.
The frequency factor (or pre-exponential factor) (A) takes into account the frequency of collisions with proper orientations.

Applications and skills – how well can you do all of the following?
Analysing graphical representation of the Arrhenius equation in its linear form $\ln k = - (E_a/RT) + \ln A$
Using the Arrhenius equation $k = Ae^{(-E_a/RT)}$.
Describing the relationships between temperature and rate constant; frequency factor and complexity of molecules colliding.
Determining and evaluating values of activation energy and frequency factors from data.

Arrhenius equation
The Arrhenius equation links the rate constant of a reaction with the temperature and activation energy. Therefore, the relationship between the rate of reaction, temperature and the presence of a catalyst can be quantified numerically. These can also be analysed graphically. The Arrhenius equation is derived from the following:

$$k = p \times Z \times e^{\frac{E_a}{RT}}$$

k = rate constant
e = natural logarithm base = 2.718...
R = universal gas constant = 8.31 J K^{-1} mol^{-1}
p = steric factor (fraction of collision with correct orientation)
Z = collision number (frequency of collision constant)
E_a = activation energy
T = temperature
p is a probability function and is independent of temperature.
Z increases with temperature but is only a minor factor.
Therefore, $p \times Z$ can be approximated to a constant.

The Arrhenius equation is then given as:

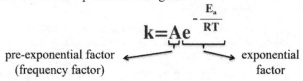

pre-exponential factor (frequency factor) — exponential factor

The pre-exponential factor represents the possible number of collisions that **could** lead to successful reactions. This is characteristic of the specific reaction.

The exponential factor represents the fraction of collisions that **would** lead to successful reactions. This is dependent on the temperature and activation energy.

By taking the natural logarithm, the Arrhenius equation can be rearranged as:

$$\ln k = -\frac{E_a}{RT} + \ln A$$

Increasing the temperature makes E_a/RT smaller. ln k increases, so the rate increases.

Adding a catalyst makes E_a and E_a/RT smaller. ln k increases, so the rate increases.

By comparing the Arrhenius equation with y = mx + c, a plot of ln k against 1/T would give a straight line with negative gradient.

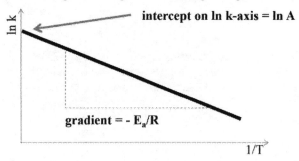

7.1 Equilibrium
Syllabus
Nature of science – can you relate this topic to these concepts?
Obtaining evidence for scientific theories—isotopic labelling and its use in defining equilibrium. Common language across different disciplines—the term dynamic equilibrium is used in other contexts, but not necessarily with the chemistry definition in mind.

Understandings – how well can you explain these statements?
A state of equilibrium is reached in a closed system when the rates of the forward and reverse reactions are equal.

The equilibrium law describes how the equilibrium constant (K_c) can be determined for a particular chemical reaction.

The magnitude of the equilibrium constant indicates the extent of a reaction at equilibrium and is temperature dependent.

The reaction quotient (Q) measures the relative amount of products and reactants present during a reaction at a particular point in time. Q is the equilibrium expression with non-equilibrium concentrations. The position of the equilibrium changes with changes in concentration, pressure, and temperature.

A catalyst has no effect on the position of equilibrium or the equilibrium constant.

Applications and skills – how well can you do all of the following?
The characteristics of chemical and physical systems in a state of equilibrium.

Deduction of the equilibrium constant expression (K_c) from an equation for a homogeneous reaction.

Determination of the relationship between different equilibrium constants (K_c) for the same reaction at the same temperature.

Application of Le Châtelier's principle to predict the qualitative effects of changes of temperature, pressure and concentration on the position of equilibrium and on the value of the equilibrium constant.

Dynamic Equilibrium

- Requires a closed system.
- In a closed system, **matter** cannot be removed or added but **energy** can be exchanged between the system and the surroundings.
- Outward appearances appear not to change but both forward and backward reactions are progressing at the same rate.
- Macroscopic properties appear to be constant i.e. concentration of each substance remains the same.

Examples of physical systems in equilibrium

Bromine liquid and gas:

$Br_2 (l) \rightleftharpoons Br_2 (g)$

Bromine liquid molecules evaporate into gaseous molecules at the same rate as the gaseous molecules are condensing into liquid molecules.

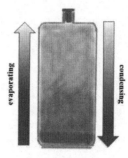

Iodine solid and gas:

$I_2 (s) \rightleftharpoons I_2 (g)$

Iodine solid molecules sublime into gaseous molecules at the same rate as the gaseous molecules are solidifying into solid molecules.

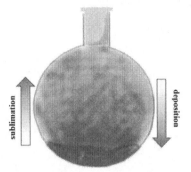

Note that in both case, **no chemical change (reaction)** occurs.

Examples of solutions in equilibrium

In a saturated solution, the rate of dissolving equals the rate of precipitation.

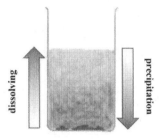

Reversible Reactions

These reactions can go in either direction at the same temperature e.g.

$H_2 (g) + I_2 (g) \rightleftharpoons 2HI (g)$

Hydrogen gas and hydrogen iodide gas are both colourless. Iodine vapour is purple.

The purple colour of the mixture fades into a lighter colour. Eventually, the colour of the reaction mixture become constant throughout the reaction vessel.

This is because the rate of the forward reaction = the rate of the reverse reaction; the system is at equilibrium.

The concentrations of H_2, I_2 and HI are constant (but not necessarily equal) at equilibrium.

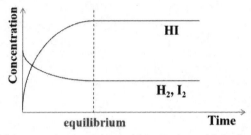

The equilibrium can also be achieved by starting with hydrogen iodide gas.

The equilibrium law

Consider a typical reversible reaction: $aA + bB \rightleftharpoons cC + dD$
The equilibrium constant can be represented by

$$K_c = \frac{[C]^c[D]^d}{[A]^a[B]^b}$$

K_c is the **equilibrium constant** with respect to concentration. It does not have any units.
Pure solids (s) and liquids (l) that appear in the chemical equation are **not included** in the expression for K_c.
Example:

For the reaction, $2H_2\ (g) + N_2\ (g) \rightleftharpoons N_2H_4\ (g)$
The equilibrium constant is

$$K_c = \frac{[N_2H_4]}{[H_2]^2[N_2]}$$

State the expression and the units for K_c for this reaction:

$Br_2\ (aq) + 2\ Fe^{2+}\ (aq) \rightleftharpoons 2Br^-\ (aq) + 2Fe^{3+}\ (aq)$

$$K_c = \frac{[Br^-]^2[Fe^{3+}]^2}{[Br_2][Fe^{2+}]^2}$$

Iron (III) ions form a distinctive deep red complex with thiocyanate ions.

$Fe^{3+}\ (aq) + NCS^-\ (aq) \rightleftharpoons Fe(NCS)^{2+}\ (aq)$
State the expression and the units for K_c for this reaction:

$$K_c = \frac{[Fe(NCS)^{2+}]}{[Fe^{3+}][NCS^-]}$$

Silver ions react with iron (II) ions in a reversible reaction:

$Ag^+\ (aq) + Fe^{2+}\ (aq) \rightleftharpoons Ag\ (s) + Fe^{3+}\ (aq)$
Define the equilibrium constant for this reaction and state its units.
N.B. This is not a homogeneous system.

$$K_c = \frac{[Fe^{3+}\ (aq)]}{[Ag^+\ (aq)]\ [Fe^{2+}\ (aq)]}$$

Le Châtelier's principle

"When a system at equilibrium is subjected to a change, it will respond in such a way as to remove the effect of the change and return the system to equilibrium."

Equilibrium position

This refers to the relative quantities of reactants and products in the equilibrium mixture. Just because the system is at equilibrium, it does not mean there are 50% reactants and 50% products. The percentage in the equilibrium mixture can (and does) vary enormously.

- If the concentration of products in the equilibrium mixture is higher, we say that the **equilibrium position** lies to the **right** of the equation.
- If the concentration of reactants in the equilibrium mixture is higher, we say that the **equilibrium position** lies to the **left** of the equation.
- If the percentage of products in the equilibrium mixture increases, we say that the **equilibrium position** has **shifted to the right**.
- If the percentage of reactants in the equilibrium mixture increases, we say that the **equilibrium position** has **shifted to the left**.

Le Châtelier's principle: Effect of concentration

If we increase the concentration of reactants
The rate of the forward reaction will increase. More products will form until equilibrium is restored. The new equilibrium mixture will have a higher proportion of products and the new equilibrium position will lie towards the right.

If we increase the concentration of products
The rate of the reverse reaction will increase. More reactants will form until equilibrium is restored. The new equilibrium mixture will have a higher proportion of reactants and the new equilibrium position will lie towards the left.

Effect on K_c
None

Example 1:

$2SO_2 \text{ (g)} + O_2 \text{ (g)} \rightleftharpoons 2SO_3 \text{ (g)}$

- Increasing the concentration of oxygen increases the rate of the forward reaction.
- More sulfur trioxide will be formed.
- The new equilibrium mixture will have a larger proportion of sulfur trioxide.

Example 2:

$Cr_2O_7^{2-} \text{ (aq)} + H_2O \text{ (l)} \rightleftharpoons 2CrO_4^{2-} \text{ (aq)} + 2H^+ \text{ (aq)}$
$Cr_2O_7^{2-}$: orange
$2CrO_4^{2-}$: yellow

- Adding NaOH (aq) will reduce the concentration of H^+ ions.
- The rate of the forward reaction increases.
- The colour of the solution changes from orange to yellow as more CrO_4^{2-} ions are formed from $Cr_2O_7^{2-}$ ions.

Le Châtelier's principle: Effect of pressure

This only applies to systems involving gases.
The pressure law states that the pressure in a closed system is inversely proportional to its volume.
Increasing the pressure will shift the equilibrium position to whichever side has the fewest molecules in order to reduce the pressure.

Effect on K_c
None

Example 1:

$$N_2 (g) + 3H_2 (g) \rightleftharpoons 2NH_3 (g)$$

- Increasing the pressure causes the equilibrium position to shift to the right (2 molecules of gas compared with 4 molecules of gas on left).
- More ammonia will be formed and the new equilibrium position will have a higher proportion of ammonia.

Example 2:

$$H_2 (g) + I_2 (g) \rightleftharpoons 2HI (g)$$

- There are equal molecules of gas on both sides of the equation.
- Changing the pressure has no effect on the equilibrium position of this system. This is because the rates of the forward and backward reactions are increased by the same amount.

Le Châtelier's principle: Effect of temperature

Increasing the temperature shifts the equilibrium position in the direction of the endothermic reaction.
The system reacts to remove the heat energy by converting it to chemical energy. This brings the temperature of the system down.

Effect on K_c
K_c is dependent on the temperature. The direction of change depends on whether the reaction is exothermic or endothermic.

If the forward reaction is endothermic:
$$CaCO_3 \text{ (s)} \rightleftharpoons CaO \text{ (s)} + CO_2 \text{ (g)} \quad \Delta H \text{ positive}$$
Increasing the temperature:
- This causes the equilibrium position to shift to the right.
- This leads to an increase in the proportion of CaO and CO_2 in the equilibrium mixture.
- The value of K_c increases.

Decreasing the temperature
- This causes the equilibrium position to shift to the left.
- This leads to an increase in the proportion of $CaCO_3$ in the equilibrium mixture.
- The value of K_c decreases.

If the forward reaction is exothermic (backward reaction is endothermic):
$$2SO_2 \text{ (g)} + O_2 \text{ (g)} \rightleftharpoons 2SO_3 \text{ (g)}$$
ΔH negative
Increasing the temperature
- This causes the equilibrium position to shift to the left.
- This leads to an increase in the proportion of SO_2 and O_2 in the equilibrium mixture.
- The value of K_c decreases.

Decreasing the temperature
- This causes the equilibrium position to shift to the right.
- This leads to an increase in the proportion of SO_3 in the equilibrium mixture.
- The value of K_c increases.

Example 1:

$N_2O_4 (g) \rightleftharpoons 2NO_2 (g)$
ΔH = +58 kJ mol⁻¹
The forward reaction is endothermic.
Increasing the temperature
- This will shift the position of equilibrium to the right.
- More dinitrogen tetroxide will decompose into nitrogen dioxide.
- The value of K_c increases.

Decreasing the temperature
- This will shift the position of equilibrium to the left.
- More dinitrogen tetroxide will be formed.
- The value of K_c decreases.

Example 2:

$N_2 (g) + 3H_2 (g) \rightleftharpoons 2NH_3 (g)$
ΔH = -92.4 kJ mol⁻¹
The forward reaction is exothermic, therefore the backward reaction is endothermic.
Increasing the temperature
- This will shift the position of equilibrium to the left.
- The yield of ammonia decreases.
- The value of K_c decreases.

Decreasing the temperature
- This will shift the position of equilibrium to the right.
- The yield of ammonia increases.
- The value of K_c increases.

Le Châtelier's principle: Effect of a catalyst

The use of a catalyst has **no effect** on the equilibrium position, it only speeds up the **rate** of the reaction.
The forward and backward reaction is equally catalysed so equilibrium is reached sooner.

Effect on K_c
None

Reaction quotient

The reaction quotient (Q) determines if a system is at equilibrium and is a measure of the relative amounts of reactants and products at a particular point in time.

$Q > K_c$	Concentration of products is greater than the equilibrium concentrations. Reverse reaction is favoured to restore equilibrium.
$Q = K_c$	System is at dynamic equilibrium. Forward reaction occurs at the same rate as the reverse reaction.
$Q < K_c$	Concentration of reactants is greater than the equilibrium concentrations. Forward reaction is favoured to restore equilibrium.

17.1 The equilibrium law
Syllabus
Nature of science – can you relate this topic to these concepts?
Employing quantitative reasoning—experimentally determined rate expressions for forward and backward reactions can be deduced directly from the stoichiometric equations and allow Le Châtelier's principle to be applied.
Understandings – how well can you explain these statements?
Le Châtelier's principle for changes in concentration can be explained by the equilibrium law.
The position of equilibrium corresponds to a maximum value of entropy and a minimum in the value of the Gibbs free energy.
The Gibbs free energy change of a reaction and the equilibrium constant can both be used to measure the position of an equilibrium reaction and are related by the equation, $\Delta G = -RT \ln K$.
Applications and skills – how well can you do all of the following?
Solution of homogeneous equilibrium problems using the expression for K_c.
Relationship between ΔG and the equilibrium constant.
Calculations using the equation $\Delta G = -RT \ln K$.

Determining the equilibrium constant (the ICE method)

I initial concentrations	Deduce the stoichiometric equation for the system. Unless specified, the initial concentrations of the products are usually zero.
C change in concentrations	Work out the change in concentration from the given data using the coefficients from the stoichiometric equation.
E equilibrium concentrations	Work out the equilibrium concentrations (difference between initial concentration and change in concentration). Substitute these values into the expression for K_c.

Note that the data given are concentrations – if amounts are given, then the concentration must be calculated using concentration = amount / volume.

Example 1

1.0 mol of phosphorus pentachloride was added to a 20 dm^3 flask and heated to 180°C. At equilibrium, 32% of the PCl$_5$ had decomposed. Calculate the value of K$_c$ at 180°C.

$$PCl_5 (g) \rightleftharpoons PCl_3 (g) + Cl_2 (g)$$

$$K_c = \frac{[PCl_3][Cl_2]}{[PCl_5]}$$

	PCl$_5$	PCl$_3$	Cl$_2$
Initial amount (mol)	1.0	0	0
Change (mol)	- 0.32	+ 0.32	+ 0.32
Equilibrium amount (mol)	1.0 – 0.32 = 0.68	0.32	0.32
Equilibrium concentration (mol dm^{-3})	0.68 / 20 = 0.034	0.32 / 20 = 0.016	0.32 / 20 = 0.016

$$K_c = \frac{0.016 \times 0.016}{0.034} = 0.0075$$

Example 2

0.40 mol of hydrogen chloride and 0.10 mol of oxygen were added to a 4.0 dm^3 flask and heated to 400°C. At equilibrium, 0.040 mol of hydrogen chloride remained. Calculate the value of K$_c$ at 400°C.

$$4HCl (g) + O_2 (g) \rightleftharpoons 2Cl_2 (g) + 2H_2O (g)$$

$$K_c = \frac{[Cl_2]^2[H_2O]^2}{[HCl]^4[O_2]}$$

	4HCl	O$_2$	2Cl$_2$	2H$_2$O
Initial amount (mol)	0.40	0.10	0	0
Change (mol)	- 0.36	- ¼ (0.36) = -0.09	+ ½ (0.36) = + 0.18	+ ½ (0.36) = + 0.18
Eq. amount (mol)	0.40 – 0.36 = 0.04	0.10 – 0.09 = 0.01	0.18	0.18
Eq. conc. (mol dm^{-3})	0.04 / 4.0 = 0.010	0.01 / 4.0 = 0.0025	0.18 / 4.0 = 0.045	0.18 / 4.0 = 0.045

$$K_c = \frac{0.045^2 \times 0.045^2}{0.010^4 \times 0.025} = 1.64 \times 10^5$$

Entropy and equilibrium

The value of the equilibrium constant, K_c gives an indication of the extent of the reaction at equilibrium. It therefore tells us whether the reactants or products are favoured at equilibrium.

The position of equilibrium occurs at:
- maximum entropy (most positive).
- minimum Gibbs free energy change (most negative).

K (or K_c) is linked to $\Delta G^\ominus$ by the following equation:
$$\Delta G^\ominus = -RT \ln K$$

By rearranging this equation, K can be calculated as follows:
$$K = e^{-\frac{\Delta G^\theta}{RT}}$$

As R is the gas constant, it can be seen that K is a constant value only if the value of T (the temperature) doesn't change. Therefore, the value of the equilibrium constant must always be quoted at a specified temperature.

By analysing the value of K from the equation, it can be summarised as follows:

K	Composition of equilibrium mixture	Gibbs free energy change, $\Delta G^\ominus$
1	neither reactans or products favoured	0
> 1	products favoured	< 0
< 1	reactants favoured	> 0

8.1 Theories of acids and bases
Syllabus
Nature of science – can you relate this topic to these concepts?
Falsification of theories—HCN altering the theory that oxygen was the element which gave a compound its acidic properties allowed for other acid–base theories to develop. Theories being superseded—one early theory of acidity derived from the sensation of a sour taste, but this had been proven false. Public understanding of science—outside of the arena of chemistry, decisions are sometimes referred to as "acid test" or "litmus test".

Understandings – how well can you explain these statements?
A Brønsted–Lowry acid is a proton/H^+ donor and a Brønsted–Lowry base is a proton/H^+ acceptor.
Amphiprotic species can act as both Brønsted–Lowry acids and bases.
A pair of species differing by a single proton is called a conjugate acid-base pair.

Applications and skills – how well can you do all of the following?
Deduction of the Brønsted–Lowry acid and base in a chemical reaction.
Deduction of the conjugate acid or conjugate base in a chemical reaction.

Preliminary concepts
Acids: substances which usually have a sour taste.
Bases: substances which react with acids to neutralise them.
Alkali: bases which are soluble in water.

Arrhenius Theory

Acids: substances which ionize in water to give hydrogen ions (H^+).
Alkalis: substances which ionize in water to give hydroxide ions (OH^-).

The neutralisation reaction between an acid and an alkali is essentially a H^+ ion reacting with a OH^- ion to from water e.g.
HCl (aq) + KOH (aq) → KCl (aq) + H_2O(l)
H^+ (aq) + Cl^- (aq) + K^+ (aq) + OH^- (aq) → K^+ (aq) + Cl^- (aq) + H_2O (l)
H^+ (aq) + OH^- (aq) → H_2O (l)
$2HNO_3$ (aq) + $Mg(OH)_2$ (aq) → $Mg(NO_3)_2$ (aq) + $2H_2O$ (l)
$2H^+$ (aq) + $2NO_3^-$ (aq) + Mg^{2+} (aq) + $2OH^-$ (aq) → Mg^{2+} (aq) + $2NO_3^-$ (aq) + $2H_2O$ (l)
$2H^+$ (aq) + $2OH^-$ (aq) → $2H_2O$ (l)

Limitatons:
explains acid-base properties in systems involving aqueous solutions only.
does not account for properties of organic substances e.g. amines as bases which do not contain OH^- groups.

Brønsted-Lowry Theory

Acids: proton (H⁺) donor.
Alkalis: proton (H⁺) acceptor.

A hydronium ion (H_3O^+) is formed when a lone pair of electrons on a water molecule forms a coordinate bond with a proton.
$H_2O + H^+ \rightarrow H_3O^+$

In any acid-base reaction there are always 2 acids and 2 bases (one on each side of the equation). The acid on one side is formed from the base on the other and vice versa.

B-L acid + B-L base ⇌ conjugate acid + conjugate base

$$HA + B \rightleftharpoons BH^+ + A^-$$

Examples:

1. **$HCl + NH_3 \rightleftharpoons NH_4^+ + Cl^-$**
   acid: HCl
   base: NH_3
   conjugate acid: NH_4^+
   conjugate base: Cl^-

2. **$NH_3 + H_2O \rightleftharpoons NH_4^+ + OH^-$**
   acid: H_2O
   base: NH_3
   conjugate acid: NH_4^+
   conjugate base: OH^-

3. **$HCl + H_2O \rightleftharpoons H_3O^+ + Cl^-$**
   acid: HCl
   base: H_2O
   conjugate acid: H_3O^+
   conjugate base: Cl^-

From examples 2 and 3, is can be seen that water is able to behave as both an acid and a base, depending upon the reaction. It is **amphiprotic**.

Note: hydrogen ions exist in their hydrated form – hydroxonium/oxonium/hydronium ions (H_3O^+). For simplicity, they are usually written in equations simply as H^+.

Common conjugate pairs

Acid	Conjugate base
HCl	Cl^-
H_2SO_4	HSO_4^-
H_2O	OH^-
CH_3COOH	CH_3COO^-
NH_4^+	NH_3

Base	Conjugate acid
OH^-	H_2O
H_2O	H_3O^+
NH_3	NH_4^+
$C_2H_5NH_2$	$C_2H_5NH_3^+$
HSO_4^-	H_2SO_4

Polyprotic acids

Monoprotic: dissociates to produce **one proton per molecule** e.g.
HCl (aq) + H_2O (l) → H_3O^+ (aq) + Cl^- (aq)
CH_3COOH (aq) + H_2O (l) → CH_3COO^- (aq) + H_3O^+ (aq)

Diprotic: dissociates to produce **two protons per molecule** e.g.
H_2SO_4 (aq) + H_2O (aq) → $2H_3O^+$ (aq) + SO_4^{2-} (aq)

Triprotic: dissociates to produce **three protons per molecule** e.g.
H_3PO_4 (aq) + H_2O (l) ⇌ $3H_2O^+$ (aq) + PO_4^{3-} (aq)

Limitaton: only describes reactions involving proton transfer.

8.2 Properties of acids and bases
Syllabus
Nature of science – can you relate this topic to these concepts?
Obtaining evidence for theories—observable properties of acids and bases have led to the modification of acid–base theories.
Understandings – how well can you explain these statements?
Most acids have observable characteristic chemical reactions with reactive metals, metal oxides, metal hydroxides, hydrogen carbonates and carbonates.
Salt and water are produced in exothermic neutralization reactions.
Applications and skills – how well can you do all of the following?
Balancing chemical equations for the reaction of acids.
Identification of the acid and base needed to make different salts.
Candidates should have experience of acid-base titrations with different indicators.

General properties of acids

Do not conduct electricity in pure form – dissolves in water to form **electrolytes**.
They form solutions with a pH < 7.0.
Changes the colour of **indicators** e.g. they turn blue litmus red.

Reaction with:
Metals
acid + metal → salt + hydrogen
Metal oxides
acid + metal oxide → salt + water
Metal carbonates
acid + metal carbonate → salt + carbon dioxide + water
Metal hydrogencarbonates
acid + metal hydrogencarbonate → metal salt + carbon dioxide + water

General properties of bases

Do not conduct electricity in pure form – some bases dissolve in water to form **electrolytes**. These bases are called **alkalis**.
Soluble bases form solutions with a pH > 7.0.
Changes the colour of **indicators** e.g. they turn red litmus blue.
Dissolve in water to release hydroxide ions, OH⁻ e.g.

$Li_2O\ (s) + H_2O\ (l) \rightarrow 2Li^+ + 2OH^-\ (aq)$

$NH_3\ (aq) + H_2O\ (l) \rightleftharpoons NH_4^+\ (aq) + OH^-\ (aq)$

$CO_3^{2-}\ (aq) + H_2O\ (l) \rightleftharpoons HCO_3^-\ (aq) + OH^-\ (aq)$

$HCO_3^-\ (aq) \rightleftharpoons CO_2\ (g) + OH^-\ (aq)$

Indicators

Indicators are used to identify substances as acids or alkalis.

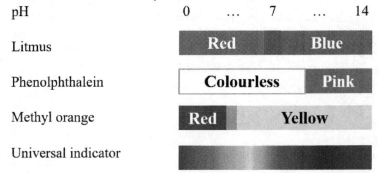

Universal indicator is a mixture of different indicators to give a wide range of colours which could be used to **deduce the pH** of a substance.

Acids and reactive metals

Acids react with **reactive metals** (those above hydrogen in the reactivity series) to give **a salt** and **hydrogen gas**.
2Li (s) + 2HCl (aq) → 2LiCl (aq) + H$_2$ (g)
lithium + hydrochloric acid → lithium chloride + hydrogen
H$_2$SO$_4$ (aq) + Zn (s) → ZnSO$_4$ (aq) + H$_2$ (g)
sulfuric acid + zinc → zinc sulfate + hydrogen
2HCl (aq) + Mg (s) → MgCl$_2$ (aq) + H$_2$ (g)
hydrochloric acid + magnesium → magnesium chloride + hydrogen
Ca (s) + 2CH$_3$COOH (aq) → (CH$_3$COO)$_2$Ca (aq) + H$_2$ (g)
calcium + ethanoic acid → calcium ethanoate + hydrogen

Acids and reactive metals (ionic equations)
The reactions can also be shown as **ionic equations** by cancelling out the **spectator ions** (ions which do not react).
2Li (s) + 2HCl (aq) → 2LiCl (aq) + H$_2$ (g)
1. Separate (aq) compounds into ions:
2Li (s) + 2H$^+$ (aq) + 2Cl$^-$ (aq) → 2Li$^+$ (aq) + 2Cl$^-$ (aq) + H$_2$ (g)
2. Cancel out spectator ions (same on both sides:
2Li (s) + 2H$^+$ (aq) + ~~2Cl$^-$ (aq)~~ → 2Li$^+$ (aq) + ~~2Cl$^-$ (aq)~~ + H$_2$ (g)
Ionic equation: **2Li (s) + 2H$^+$ (aq) → 2Li$^+$ (aq) + H$_2$ (g)**

Acids and bases

Acids react with bases, such as **metal oxides** and **hydroxides** to form **a salt** and **water** in a **neutralisation** reaction.

$2HNO_3$ (aq) + CuO (s) → $Cu(NO_3)_2$ (aq) + H_2O (l)
nitric acid + copper(II) oxide → copper(II) nitrate + water

$2HCl$ (aq) + CaO (s) → $CaCl_2$ (aq) + H_2O (l)
hydrochloric acid + calcium oxide → calcium chloride + water

H_2SO_4 (aq) + $2NH_4OH$ (aq) → $(NH_4)_2SO_4$ (aq) + $2H_2O$ (l)
sulfuric acid + ammonium hydroxide → ammonium sulfate + water

H_3PO_4 (aq) + $3NaOH$ (aq) → Na_3PO_4 (aq) + $3H_2O$ (l)
phosphoric acid + sodium hydroxide → sodium phosphate + water

Acids and bases (ionic equations)

The reactions between acids and aqueous alkalis can all be summarised by the **same ionic equation**.

H_2SO_4 (aq) + $2NH_4OH$ (aq) → $(NH_4)_2SO_4$ (aq) + $2H_2O$ (l)

$2H^+$ (aq) + ~~SO_4^{2-} (aq) + $2NH_4^+$ (aq)~~ + $2OH^-$ (aq) → ~~$2NH_4^+$ (aq) + SO_4^{2-} (aq)~~ + $2H_2O$ (l)

Ionic equation: $2H^+$ (aq) + $2OH^-$ (aq) → $2H_2O$ (l)
            H^+ (aq) + OH^- (aq) → H_2O (l)

Acids and carbonates

Acids react with **carbonates** to give **a salt**, **water** and **carbon dioxide**, which appears as **effervescence** (bubbles).

$2HCl$ (aq) + $ZnCO_3$ (s) → $ZnCl_2$ (aq) + H_2O (l) + CO_2 (g)
hydrochloric acid + zinc carbonate → zinc chloride + water + carbon dioxide

H_2SO_4 (aq) + $CaCO_3$ (s) → $CaSO_4$ (aq) + H_2O (l) + CO_2 (g)
sulfuric acid + calcium carbonate → calcium sulfate + water + carbon dioxide

Acids and hydrogencarbonates

Acids react with **hydrogencarbonates** to give **a salt**, **water** and **carbon dioxide**, which appears as **effervescence** (bubbles).

HCl (aq) + $NaHCO_3$ (s) → $NaCl$ (aq) + H_2O (l) + CO_2 (g)
hydrochloric acid + sodium hydrogencarbonate → sodium chloride + water + carbon dioxide

$2CH_3COOH$(aq) + $Ca(HCO_3)_2$(s) → $(CH_3COO)_2Ca$(aq) + $2H_2O$(l) + $2CO_2$(g)
ethanoic acid + calcium hydrogencarbonate → calcium ethanoate + water + carbon dioxide

8.3 The pH scale
Syllabus
Nature of science – can you relate this topic to these concepts?
Occam's razor—the pH scale is an attempt to scale the relative acidity over a wide range of H concentrations into a very simple number.
Understandings – how well can you explain these statements?
$pH = -\log[H^+ (aq)]$ and $[H^+] = 10^{-pH}$
A change of one pH unit represents a 10-fold change in the hydrogen ion concentration $[H^+]$.
pH values distinguish between acidic, neutral and alkaline solutions.
The ionic product constant, $K_w = [H^+][OH^-] = 10^{-14}$ at 298 K.
Applications and skills – how well can you do all of the following?
Solving problems involving pH, $[H^+]$ and $[OH^-]$.
Students should be familiar with the use of a pH meter and universal indicator.

pH facts
- a measure of acidity (or alkalinity) of a solution.
- usually on a scale from 0 to 14 (concentrated solutions of strong acids and bases it can extend beyond this range).
- at 25°C, the pH of pure water is 7 (neutral).
- pH < 7: the solution is acidic (there is an excess of H^+).
- pH > 7: the solution is alkaline (there is an excess of OH^-).

The pH scale in context

pH	0	4	7	10	14
$[H^+ (aq)]$	10^{-0}	10^{-4}	10^{-7}	10^{-10}	10^{-14}
$[OH^- (aq)]$	10^{-14}	10^{-10}	10^{-7}	10^{-4}	10^{-0}
Universal indicator colour	Red	Orange	Green	Blue	Purple
Acidity / Basicity	Strong acid	Weak acid	Neutral	Weak base	Strong base
Examples	Dilute strong acid	Vinegar, acid rain	Pure water	Antacid solution, toothpaste, detergent	Dilute strong alkali, caustic soda, oven cleaner

Measuring pH

The pH of a solution can be measure by using a pH meter. This is an electrode that measures the [H$^+$ (aq)] through a special electrode (pH probe) and outputs the data to a digital display but requires:
- calibration with a buffer solution.
- standardisation at specific temperatures.

pH of strong acids

The pH of a solution is defined as the negative logarithm to the base 10 of the hydrogen ion concentration in mol dm^{-3}.

pH = - log$_{10}$ [H$^+$]
[H$^+$] = 10^{-pH}

pH is a logarithmic scale: when [H$^+$] is increased by a factor of ten, the pH decreases by one unit.
Example:
[H$^+$] = 10^{-4} mol dm^{-3} gives pH 4.0
[H$^+$] = 10^{-3} mol dm^{-3} gives pH 3.0

An **acidic** solution is defined as one in which **[H$^+$] > [OH$^-$]**.
At **25°C**,
[H$^+$] > 1.0 x 10^{-7} mol dm^{-3}
pH < 7.00

[H$^+$] / mol dm^{-3}	pH
1.00 x 10^{-5}	5.00
5.00 x 10^{-4}	3.30
2.00 x 10^{-12}	11.70

For **monoprotic** strong acids, the concentration of hydrogen ions is the same as the initial concentration of the acid i.e. **[H$^+$] = [acid]** e.g.
What is the pH of hydrochloric acid of concentration 0.0025 mol dm^{-3}?
pH = - log (0.0025) = 2.6

Diprotic acids e.g. H$_2$SO$_4$ can potentially produce two H$^+$ ions per molecule i.e. **[H$^+$] = 2 x [acid]** e.g.
What is the pH of sulfuric acid of concentration 0.0025 mol dm^{-3}?
[H$^+$] = 2 x 0.0025 = 0.0050 mol dm^{-3}
pH = - log (0.0050) = 2.30

Neutrality of pure water

Water dissociates to a very slight degree:

$H_2O \ (l) \rightleftharpoons H^+ \ (aq) + OH^- \ (aq)$

Pure water is neutral because [H$^+$] and [OH$^-$] from dissociation are equal.

At 298K, both [H$^+$] and [OH$^-$] have values of 10^{-7} mol dm^{-3}. The equilibrium above can be expressed as:

$$K_c = \frac{[H^+][OH^-]}{[H_2O]}$$

Ionic product constant

The degree of **auto-ionisation** is very small therefore [H$_2$O] is assumed to be constant.

The **ionic product constant** for water (K$_w$) is defined as:

K$_w$ = [H$^+$][OH$^-$] = 1.0 x 10^{-7} x 1.0 x 10^{-7} = 1.0 x 10^{-14} at 298K

As K$_w$ has been derived from the equilibrium constant K$_c$ above, it is essentially also an expression of equilibrium. Therefore, the value for K$_w$ must always be quoted with the specific temperature.

pH of strong alkalis

An **alkaline** solution is defined as one in which $[H^+] < [OH^-]$.
At **25°C**,
$[H^+] < 1.0 \times 10^{-7}$ mol dm^{-3}
pH > 7.00
To calculate the pH of an alkaline solution, follow these steps:
- **Convert $[OH^-]$ to $[H^+]$ by using the using the formula $K_w = [H^+][OH^-]$**
- **Calculate pH from the equation pH = -log $[H^+]$**

Example:

Find the pH of a solution where $[OH^-] = 2.20 \times 10^{-2}$ mol dm^{-3}
$[H^+][OH^-] = 1.0 \times 10^{-14}$
$[H^+] = (1.0 \times 10^{-14})/[OH^-] = (1.0 \times 10^{-14})/(2.20 \times 10^{-2}) = 4.6 \times 10^{-13}$
pH = - log $[H^+]$ = - log (4.6×10^{-13}) = 12.3

Question 1

Find the pH of a 2.00 mol dm^{-3} solution of sodium hydroxide.
NaOH + aq → Na$^+$ (aq) + OH$^-$ (aq)
$[H^+][OH^-] = 1.0 \times 10^{-14}$
$[H^+] = (1.0 \times 10^{-14})/[OH^-] = (1.0 \times 10^{-14})/(2.00) = 5.0 \times 10^{-15}$
pH = - log $[H^+]$ = - log (5.0×10^{-15}) = 14.3

Question 2

Find the pH of a 0.222 mol dm^{-3} solution of barium hydroxide.
Ba(OH)$_2$ + aq → Ba^{2+} (aq) + 2OH$^-$ (aq)
$[H^+][OH^-] = 1.0 \times 10^{-14}$
$[H^+] = (1.0 \times 10^{-14})/[OH^-] = (1.0 \times 10^{-14})/(0.444) = 2.3 \times 10^{-14}$
pH = - log $[H^+]$ = - log (2.3×10^{-14}) = 13.6

8.4 Strong and weak acids and bases
Syllabus
Nature of science – can you relate this topic to these concepts?
Improved instrumentation—the use of advanced analytical techniques has allowed the relative strength of different acids and bases to be quantified. Looking for trends and discrepancies—patterns and anomalies in relative strengths of acids and bases can be explained at the molecular level. The outcomes of experiments or models may be used as further evidence for a claim—data for a particular type of reaction supports the idea that weak acids exist in equilibrium.
Understandings – how well can you explain these statements?
Strong and weak acids and bases differ in the extent of ionization. Strong acids and bases of equal concentrations have higher conductivities than weak acids and bases.
A strong acid is a good proton donor and has a weak conjugate base.
A strong base is a good proton acceptor and has a weak conjugate acid.
Applications and skills – how well can you do all of the following?
Distinction between strong and weak acids and bases in terms of the rates of their reactions with metals, metal oxides, metal hydroxides, metal hydrogen carbonates and metal carbonates and their electrical conductivities for solutions of equal concentrations.

Strong acids

These are acids which are **almost completely dissociated (ionised)** in dilute aqueous solution. They are very effective **proton donors**.

$$HA\ (aq) + H_2O\ (l) \rightleftharpoons H_3O^+\ (aq) + A^-\ (aq)$$
$\approx 0\%$ $\approx 100\%$ (equilibrium lies far to the **right**)

The equation is usually written with a **forward arrow** as the **backward reaction is negligible**.

Examples:
Hydrochloric acid: $HCl\ (aq) + H_2O\ (l) \rightarrow H_3O^+\ (aq) + Cl^-\ (aq)$
Hydrobromic acid: $HBr\ (aq) + H_2O\ (l) \rightarrow H_3O^+\ (aq) + Br^-\ (aq)$
Sulfuric acid: $H_2SO_4\ (aq) + H_2O\ (l) \rightarrow H_3O^+\ (aq) + HSO_4^-\ (aq)$
Nitric acid: $HNO_3\ (aq) + H_2O\ (l) \rightarrow H_3O^+\ (aq) + NO_3^-\ (aq)$

Weak acids

These are acids which **only slightly dissociate** into ions in dilute aqueous solution. They are poor **proton donors**.

$$HA (aq) + H_2O (l) \rightleftharpoons H^+ (aq) + A^- (aq)$$
$\approx 99\%$ $\qquad\qquad\qquad\qquad$ $\approx 1\%$ (**equilibrium lies far to the left**)

Almost all organic acids are weak acids.

Examples:

Ethanoic aid: $CH_3COOH (aq) + H_2O (l) \rightleftharpoons H_3O^+ (aq) + CH_3COO^- (aq)$

Phosphoric acid: $H_3PO_4 (aq) + H_2O (l) \rightleftharpoons H_3O^+ (aq) + H_2PO_4^- (aq)$

Aqueous carbon dioxide (carbonic acid) behaves as a weak acid. This can be written in two ways:

$CO_2 (aq) + 2H_2O (l) \rightleftharpoons H_3O^+ (aq) + HCO_3^- (aq)$

$H_2CO_3 (aq) + H_2O (l) \rightleftharpoons H_3O^+ (aq) + HCO_3^- (aq)$

Strong bases

These are bases which are **completely dissociated** into ions in aqueous solution. The equations are written as **one-directional** as the **backward reaction is negligible**.

Examples:
All Group 1 hydroxides e.g.
sodium hydroxide: $NaOH (aq) \rightarrow Na^+ (aq) + OH^- (aq)$
barium hydroxide: $Ba(OH)_2 (aq) \rightarrow Ba^{2+} (aq) + 2OH^- (aq)$

Weak bases
These are bases which **only slightly dissociate** into ions in dilute aqueous solution. An **equilibrium** exists between the base and the hydroxide ions.

Examples:

Ammonia: $NH_3\ (aq) + H_2O\ (l) \rightleftharpoons NH_4^+\ (aq) + OH^-\ (aq)$

Ethylamine: $C_2H_5NH_2\ (aq) + H_2O\ (l) \rightleftharpoons C_2H_5NH_3^+\ (aq) + OH^-\ (aq)$

The anions formed by weak acids e.g. the carbonate, ethanoate and phosphate ions also act as weak bases:

$CO_3^{2-}\ (aq) + H_2O\ (l) \rightleftharpoons HCO_3^-\ (aq) + OH^-\ (aq)$

Differentiating strong and weak acids and bases

A. **pH**
- pH is a measure of the **H⁺ (and indirectly, OH⁻) concentration**.
- For solutions of equal concentration:
    - A **weak acid** has a **higher pH** than a strong acid.
    - A **strong base** has a **higher pH** than a weak base.

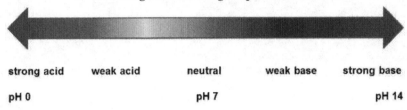

strong acid	weak acid	neutral	weak base	strong base
pH 0		pH 7		pH 14

B. **Electrical conductivity**
- In solution, the **electrical charge carriers** are the **mobile positive ions (cations) and negative ions (anions)**.
- The **degree of dissociation** of acids and bases can be measured by the **level of electrical conductivity** of the solutions.
- For solutions of equal concentration:
    - Weak acids and alkalis do not conduct electricity as well as strong acids and alkalis, but they conduct electricity better than water.
    - Strong acids and alkalis are more effective **electrolytes** than weak acids and alkalis.

C. **Rate of reactions**
- Weak acids react more slowly in acid reactions (such as those with a carbonate to give carbon dioxide or with an active metal to give hydrogen gas) than strong acids of equal concentration.
- The relative strength of the acids can be **qualitatively** determined by the **rate of effervescence**.

Differentiating between 'strength' and 'concentration'

The terms strong/weak and concentrated/dilute have very different meanings.

HCl (aq) 0.100 mol dm^{-3}
pH = -log (0.100) = 1.00 (strong and more concentrated)
CH$_3$COOH (aq) 0.100 mol dm^{-3} (a weak acid with an assumed 1% degree of dissociation)
pH = -log (0.100 x 1/100) = 3.00 (weak and more concentrated)
HCl (aq) 0.001 cmol dm^{-3}
pH = -log (0.001) = 3.00 (strong and more dilute)

Enthalpy change of neutralisation (ΔH_{neut})

Strong acids and bases are already fully dissociated in solution. The enthalpy change of neutralisation is due only to the reaction between H$^+$ and OH$^-$ ions.

HCl (aq) + KOH (aq) → KCl (aq) + H$_2$O (l) ΔH_{neut} = -57.1 kJ mol^{-1}
H$_2$SO$_4$ (aq) + 2NaOH (aq) → Na$_2$SO$_4$ (aq) + 2H$_2$O (l) ΔH_{neut} = -114.2 kJ mol^{-1}

Weak acids or bases remain mostly undissociated in solution. Neutralisation reactions involving weak acids or bases will therefore require the species to dissociate in order to provide the H$^+$ and OH$^-$ ions for reaction. This process is slightly endothermic.
CH$_3$COOH (aq) + H$_2$O (l) ⇌ CH$_3$COO$^-$ (aq) + H$_3$O$^+$ (aq) ΔH_{dis} = +1.0 kJ mol^{-1}
CH$_3$COOH (aq) + NaOH (aq) → CH$_3$COONa (aq) + H$_2$O (l) ΔH_{neut} = -56.1 kJ mol^{-1}
NH$_3$ (aq) + H$_2$O (l) ⇌ NH$_4^+$ (aq) + OH$^-$ (aq) ΔH_{dis} = +3.7 kJ mol^{-1}
NH$_3$ (aq) + HCl (aq) → NH$_4$Cl (aq) + H$_2$O ΔH_{neut} = -53.4 kJ mol^{-1}
CH$_3$COOH (aq) + NH$_3$ (aq) → CH$_3$COONH$_4$ (aq) ΔH_{neut} = -50.4 kJ mol^{-1}
theoretical value = -52.7 kJ mol^{-1}

8.5 Acid deposition
Syllabus
Nature of science – can you relate this topic to these concepts?
Risks and problems—oxides of metals and non-metals can be characterized by their acid–base properties. Acid deposition is a topic that can be discussed from different perspectives. Chemistry allows us to understand and to reduce the environmental impact of human activities.

Understandings – how well can you explain these statements?
Rain is naturally acidic because of dissolved CO_2 and has a pH of 5.6. Acid deposition has a pH below 5.6.

Acid deposition is formed when nitrogen or sulfur oxides dissolve in water to form HNO_3, HNO_2, H_2SO_4 and H_2SO_3.

Sources of the oxides of sulfur and nitrogen and the effects of acid deposition should be covered.

Applications and skills – how well can you do all of the following?
Balancing the equations that describe the combustion of sulfur and nitrogen to their oxides and the subsequent formation of H_2SO_3, H_2SO_4, HNO_2 and HNO_3.

Distinction between the pre-combustion and post-combustion methods of reducing sulfur oxides emissions.

Deduction of acid deposition equations for acid deposition with reactive metals and carbonates.

Definitions

Acid deposition: The process of acid forming pollutants being deposited onto the Earth's surface.
Acid rain: The reduced pH of rainwater due to the dissolving of non-metal oxide pollutants e.g. sulfur oxides and nitrogen oxides in rainwater.

Natural rainwater

Natural rainwater is slightly acidic (pH ≈ 5.6) due to the presence of atmospheric CO_2 which dissolves to form carbonic acid, H_2CO_3 which is a weak acid.
CO_2 (g) + H_2O (l) ⇌ H_2CO_3 (aq)
H_2CO_3 (aq) ⇌ H^+ (aq) + HCO_3^- (aq)
HCO_3^- (aq) ⇌ H^+ (aq) + CO_3^{2-} (aq)

Acid rain – effect of nitrogen oxides

The internal combustion engine is the main source of nitrogen oxides. The intense high temperature causes the reaction between atmospheric nitrogen and oxygen to form **nitrogen(II) oxide (nitrogen monoxide)**:
N_2 (g) + O_2 (g) → 2NO (g)
NO reacts further with oxygen in the internal combustion engine and also with atmospheric oxygen to form **nitrogen(IV) oxide (nitrogen dioxide)**:
2NO (g) + O_2 (g) → 2NO_2 (g)
NO_2 appears as a brown smog and dissolves in rainwater to form **nitric acid** and **nitrous acid**:
2NO_2 (g) + H_2O (l) → HNO_3 (aq) + HNO_2 (aq)
HNO_2 is oxidised by atmospheric oxygen to HNO_3:
HNO_2 (aq) + O_2 (g) → 2HNO_3 (aq)
Therefore, the eventual pollutant by all nitrogen oxides is **nitric acid, HNO_3**.

Acid rain – effect of sulfur oxides

The combustion of fossil fuels is the main source of sulfur oxides. Sulfur or sulfur compounds are natural impurities found in fossil fuels. On combustion, **sulfur dioxide** is formed:

$$S(s) + O_2(g) \rightarrow SO_2(g)$$

SO_2 reacts further with atmospheric oxygen to form **sulfur trioxide**:

$$2SO_2(g) + O_2(g) \rightleftharpoons 2SO_3(g)$$

SO_2 dissolves in rainwater to form **sulfurous acid** (a weak acid):

$$SO_2(g) + H_2O(l) \rightleftharpoons H_2SO_3(aq)$$
$$H_2SO_3(aq) + H_2O(l) \rightleftharpoons HSO_3^-(aq) + H_3O^+(aq)$$

SO_3 dissolves in rainwater to form **sulfuric acid** (a strong acid):

$$SO_3(g) + H_2O(l) \rightarrow H_2SO_4(aq)$$
$$H_2SO_4(aq) + 2H_2O(l) \rightarrow 2H_3O^+(aq) + SO_4^{2-}(aq)$$

Measures to reduce acid rain pollutants

Pre-combustion methods
Fossil fuels are processed (cleaned) before use by:

- **Mineral beneficiation** – separating the valuable constituents of fuel from the impurities by physical means such as gravity.
- **Flotation** – discarding the impurities contained in a fuel based on its density or solubility. Modern methods employ highly selective bacteria to aid in this process.

Post-combustion methods
Sulfur dioxide, nitrogen oxides, heavy metals and dioxins are removed from exhaust gases before they are released into the atmosphere:

- **Flue gas desulfurisation (scrubbing)** – sulfur dioxide can be removed by lining chimneys with limestone ($CaCO_3$) or lime (CaO).
$$CaO(s) + SO_2(g) \rightarrow CaSO_3(s)$$
- **Wet scrubbing** – sea water is a natural alkali which can remove SO_2 when sprayed within chimneys.

Effects of acid rain on buildings

Limestone and **marble** structures are most susceptible to acid rain due to their high content of **calcium carbonate (CaCO$_3$)**. This causes gradual erosion as the carbonate reacts with acid in a neutralisation reaction.
CO_3^{2-} (s) + $2H^+$ (aq) → CO_2 (g) + H_2O (l)

18.1 Lewis acids and bases
Syllabus
Nature of science – can you relate this topic to these concepts?
Theories can be supported, falsified or replaced by new theories – acid-theories can be extended to a wider field of applications by considering lone pairs of electrons. Lewis theory doesn't falsify Brønsted-Lowry but extends it.
Understandings – how well can you explain these statements?
A Lewis acid is a lone pair acceptor and a Lewis base is a lone pair donor.
When a Lewis base reacts with a Lewis acid a coordinate bond is formed.
A nucleophile is a Lewis base and an electrophile is a Lewis acid.
Applications and skills – how well can you do all of the following?
Application of Lewis' acid–base theory to inorganic and organic chemistry to identify the role of the reacting species.

Preliminary definitions

Nucleophile: a lone pair electron donor.
Electrophile: a lone pair electron acceptor.
In other words, an **electrophile** seeks to accept the pair of electrons donated by a **nucleophile**.
Lewis acids are **electrophiles** – they accept a pair of electrons to form a coordinate bond.
Lewis bases are **nucleophiles** – they donate a pair of electrons to form a coordinate bond.

Examples of Lewis acids and bases

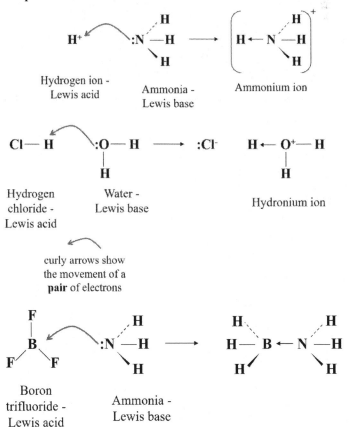

Electronic configuration around the boron atom:

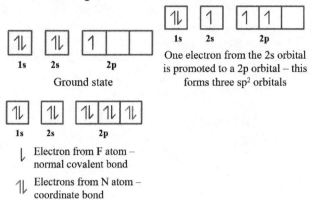

Ground state

One electron from the 2s orbital is promoted to a 2p orbital – this forms three sp² orbitals

↓ Electron from F atom – normal covalent bond

↑↓ Electrons from N atom – coordinate bond

Complex compounds: Ligands such as H_2O, NH_3 and Cl^- act as Lewis bases by donating a pair of electrons to the central ion (Lewis acid).

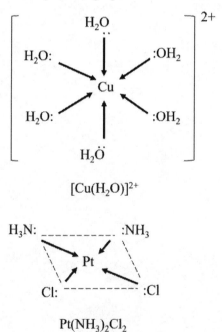

$[Cu(H_2O)]^{2+}$

$Pt(NH_3)_2Cl_2$

(cis-platin)

18.2 Calculations involving acids and bases
Syllabus
Nature of science – can you relate this topic to these concepts?
Obtaining evidence for scientific theories—application of the equilibrium law allows strengths of acids and bases to be determined and related to their molecular structure.
Understandings – how well can you explain these statements?
The expression for the dissociation constant of a weak acid (K_a) and a weak base (K_b).
For a conjugate acid base pair, $K_a \times K_b = K_w$.
The relationship between K_a and pK_a is ($pK_a = -\log K_a$), and between K_b and pK_b is ($pK_b = -\log K_b$).
Applications and skills – how well can you do all of the following?
Solution of problems involving [H^+ (aq)], [OH^- (aq)], pH, pOH, K_a, pK_a, K_b and pK_b.
Discussion of the relative strengths of acids and bases using values of K_a, pK_a, K_b and pK_b.

Quick review of pH

The pH of a solution is defined as the negative logarithm to the base 10 of the hydrogen ion concentration in mol dm^{-3}.

pH = - log [H$^+$]

A new term: pOH
The pOH of a solution is defined as the negative logarithm to the base 10 of the hydroxide ion concentration in mol dm^{-3}.

pOH = - log [OH$^-$]

Acid dissociation constant, K_a

For the dissociation reaction:

$$HA\ (aq) + H_2O\ (l) \rightleftharpoons H_3O^+\ (aq) + A^-\ (aq)$$

The degree of dissociation of weak acids is so small that $[H_2O]$ can be assumed to be constant.
$[H_3O^+]$ is usually written as a single proton, $[H^+]$.
K_a for weak acids are usually very small.
It is more convenient to use pK_a where **$pK_a = -\log K_a$**

$$K_a = \frac{[H_3O^+][A^-]}{[HA][H_2O]}$$

$$K_a = \frac{[H_3O^+][A^-]}{[HA]}$$

$$K_a = \frac{[H^+][A^-]}{[HA]}$$

Base dissociation constant, K_b

For the dissociation reaction:

$$B\ (aq) + H_2O\ (l) \rightleftharpoons BH^+\ (aq) + OH^-\ (aq)$$

The degree of dissociation of weak acids is so small that $[H_2O]$ can be assumed to be constant.
The reaction is often simplified as:

$$BOH\ (aq) \rightleftharpoons B^+\ (aq) + OH^-\ (aq)$$

K_b for weak bases are usually very small.
It is more convenient to use pK_b where **$pK_b = -\log K_b$**

$$K_b = \frac{[BH^+][OH^-]}{[B][H_2O]}$$

$$K_b = \frac{[BH^+][OH^-]}{[B]}$$

$$K_b = \frac{[B^+][OH^-]}{[BOH]}$$

Size of K_a, pK_a, K_b and pK_b

Acid	Formula	K_a / mol dm^{-3}	pK_a
sulphuric acid	H_2SO_4	very large	-
nitric acid	HNO_3	40	- 1.6
ethanoic acid	CH_3COOH	1.74 x 10^{-5}	4.76

The larger the K_a, the **stronger** the acid.
The larger the pK_a, the **weaker** the acid.

Base	Formula	K_b / mol dm^{-3}	pK_b
diethylamine	$(C_2H_5)_2NH$	6.92 x 10^{-4}	3.16
methylamine	CH_3NH_2	4.57 x 10^{-4}	3.34
ammonia	NH_3	1.78 x 10^{-5}	4.75

The larger the K_b, the **stronger** the base.
The larger the pK_b, the **weaker** the base.

pH of weak acids
Example:
Butanoic acid, C_3H_7COOH of concentration 0.005 mol dm^{-3} has a pH of 3.62. Calculate the K_a and pK_a value of this weak acid at 298 K.
$[H^+] = 10^{-pH} = 10^{-3.62} = 2.40 \times 10^{-4}$ mol dm^{-3}

$$C_3H_7COOH\ (aq) \rightleftharpoons C_3H_7COO^-\ (aq) + H^+\ (aq)$$

	$C_3H_7COOH\ (aq)$	$C_3H_7COO^-\ (aq)$	$H^+\ (aq)$
Initial conc. (mol dm^{-3})	0.005	0.000	0.000
Change in conc. (mol dm^{-3})	-2.40 x 10^{-4}	+2.40 x 10^{-4}	+2.40 x 10^{-4}
Equilibrium conc. (mol dm^{-3})	0.005 – 2.40 x 10^{-4}	2.40 x 10^{-4}	2.40 x 10^{-4}

$$K_a = \frac{[C_3H_7COO^-][H^+]}{[C_3H_7COOH]} = \frac{(2.40 \times 10^{-4})(2.40 \times 10^{-4})}{0.005} = 1.15 \times 10^{-5}\ \text{mol dm}^{-3}$$

$[C_3H_7COOH]$ at equilibrium is taken as 0.005 because the percentage change:
$(2.40 \times 10^{-4})/0.005 \times 100 = 4.8\%$ is less than 5% (boundary for experimental error).

$$pK_a = -\log(1.15 \times 10^{-5}) = 4.94$$

Question 1:
Methanoic acid, HCOOH of concentration 0.100 mol dm^{-3} has a pH of 2.41. Calculate the K_a and pK_a value of this weak acid at 298 K.
$[H^+] = 10^{-pH} = 10^{-2.41} = 3.89 \times 10^{-3}$ mol dm^{-3}

	HCOOH (aq) $\rightleftharpoons$	HCOO$^-$ (aq) +	H$^+$ (aq)
Initial conc. (mol dm^{-3})	0.100	0.000	0.000
Change in conc. (mol dm^{-3})	-3.89 x 10^{-3}	+3.89 x 10^{-3}	+3.89 x 10^{-3}
Equilibrium conc. (mol dm^{-3})	0.100 – 3.89 x 10^{-3}	3.89 x 10^{-3}	3.89 x 10^{-3}

$$K_a = \frac{[HCOO^-][H^+]}{[HCOOH]} = \frac{3.89 \times 10^{-3} \times 3.89 \times 10^{-3}}{0.100} = 1.51 \times 10^{-4} \text{ mol dm}^{-3}$$

[HCOOH] at equilibrium is taken as 0.100 because the percentage change:
(3.89x10^{-3})/0.100 x 100 = 3.9% is less than 5% (boundary for experimental error).
$$pK_a = -\log(1.51 \times 10^{-4}) = 3.82$$

Question 2:
Hydrofluoric acid, HF of concentration 0.050 mol dm^{-3} has a pH of 2.22. Calculate the K_a and pK_a value of this weak acid at 298 K.
$[H^+] = 10^{-pH} = 10^{-2.22} = 6.03 \times 10^{-3}$ mol dm^{-3}

	HF (aq) $\rightleftharpoons$	H$^+$ (aq) +	F$^-$ (aq)
Initial conc. (mol dm^{-3})	0.050	0.000	0.000
Change in conc. (mol dm^{-3})	-6.03 x 10^{-3}	+6.03 x 10^{-3}	+6.03 x 10^{-3}
Equilibrium conc. (mol dm^{-3})	0.050 – 6.03 x 10^{-3}	6.03 x 10^{-3}	6.03 x 10^{-3}

$$K_a = \frac{[H^+][F^-]}{[HF]} = \frac{6.03 \times 10^{-3} \times 6.03 \times 10^{-3}}{0.050 - 6.03 \times 10^{-3}} = 8.27 \times 10^{-4} \text{ mol dm}^{-3}$$

[HF] at equilibrium is not approximated because the percentage change:
(6.03x10^{-3})/0.050 x 100 = 12.1% is more than 5% (boundary for experimental error).
$$pK_a = -\log(8.27 \times 10^{-4}) = 3.08$$

pH of weak bases
Example:
Dimethylamine of concentration 0.005 mol dm^{-3} has a pH of 11.21.
Calculate the K_b and pK_b value of this weak base at 298 K.
pOH = 14 − 11.21 = 2.79
[OH$^-$] = 10^{-pOH} = 10$^{-2.79}$ = 1.62 x 10^{-3} mol dm^{-3}

	(CH$_3$)$_2$NH (aq) + H$_2$O (l) ⇌	(CH$_3$)$_2$NH$_2$$^+$ (aq) +	OH$^-$ (aq)
Initial conc. (mol dm^{-3})	0.005	0.000	0.000
Change in conc. (mol dm^{-3})	-1.62 x 10^{-3}	+1.62 x 10^{-3}	+1.62 x 10^{-3}
Equilibrium conc. (mol dm^{-3})	0.005 − 1.62 x 10^{-3}	1.62 x 10^{-3}	1.62 x 10^{-3}

$$K_a = \frac{[(CH_3)_2NH_2^+][OH^-]}{[(CH_3)_2NH]} = \frac{1.62 \times 10^{-3} \times 1.62 \times 10^{-3}}{0.005 - 1.62 \times 10^{-3}} = 7.76 \times 10^{-4} \text{ mol dm}^{-3}$$

[(CH$_3$)$_2$NH] at equilibrium is not approximated because the percentage change:
(1.62x10^{-3})/0.005 x 100 = 32.4% is more than 5% (boundary for experimental error).

$$pK_a = -\log(7.76 \times 10^{-4}) = 3.11$$

Question:
Ammonia of concentration 0.100 mol dm^{-3} has a pH of 11.12.
Calculate the K_b and pK_b value of this weak base at 298 K.
pOH = 14 − 11.12 = 2.88
[OH$^-$] = 10^{-pOH} = 10$^{-2.88}$ = 1.32 x 10^{-3} mol dm^{-3}

	NH$_3$ (aq) + H$_2$O (l) ⇌	NH$_4$$^+$ (aq) +	OH$^-$ (aq)
Initial conc. (mol dm^{-3})	0.100	0.000	0.000
Change in conc. (mol dm^{-3})	-1.32 x 10^{-3}	+1.32 x 10^{-3}	+1.32 x 10^{-3}
Equilibrium conc. (mol dm^{-3})	0.100 − 1.32 x 10^{-3}	1.32 x 10^{-3}	1.32 x 10^{-3}

$$K_a = \frac{[NH_4^+][OH^-]}{[NH_3]} = \frac{1.32 \times 10^{-3} \times 1.32 \times 10^{-3}}{0.100} = 1.74 \times 10^{-5} \text{ mol dm}^{-3}$$

[NH$_3$] at equilibrium is not approximated because the percentage change:
(1.32x10^{-3})/0.100 x 100 = 1.32% is less than 5% (boundary for experimental error).

$$pK_a = -\log(1.74 \times 10^{-5}) = 4.76$$

Conjugate acid-base pairs

Acid dissociation constant, K_a

For the dissociation reaction: $HA\ (aq) \rightleftharpoons H^+\ (aq) + A^-\ (aq)$

$$K_a = \frac{[H^+][A^-]}{[HA]}$$

Base dissociation constant, K_b

The conjugate base reacts with water: $A^-\ (aq) + H_2O\ (l) \rightleftharpoons HA\ (aq) + OH^-\ (aq)$

$$K_b = \frac{[HA][OH^-]}{[A^-]}$$

Taking the product of K_a and K_b,

$$K_a \cdot K_b = \frac{[H^+][A^-]}{[HA]} \cdot \frac{[HA][OH^-]}{[A^-]} = [H^+][OH^-] = K_w$$

$$K_a = \frac{K_w}{K_b} \text{ and } K_b = \frac{K_w}{K_a}$$

Taking the logarithm: $pK_a + pK_b = pK_w$
At 25°C, $K_w = 1.00 \times 10^{-14}$
$pK_w = 14.00$
$pK_a + pK_b = 14.00$
This relationship reveals the following trends:
- a higher value of pK_a gives a lower value of pK_b.
- stronger acids have conjugate bases which are weaker.
- stronger bases have conjugate acids which are weaker.

Ionic product of water, K_w (review)

$H_2O\ (l) \rightleftharpoons H^+\ (aq) + OH^-\ (aq)$ $\Delta H = +57$ kJ mol^{-1}
In pure water at 25°C, $[H^+] = [OH^-] = 1.0 \times 10^{-7}$ mol dm^{-3}
$K_w = [H^+][OH^-] = 1.0 \times 10^{-14}$ at 298 K
As the forward reaction is endothermic, increasing the temperature:
- shifts the equilibrium to the right.
- increases the equilibrium constant.

Effect of temperature rise on pH
$[H^+] > 10^{-7}$ mol dm^{-3}, so the pH of pure water is < 7.
Effect of temperature rise on K$_w$
At temperatures above 298 K, pure water is still neutral i.e. $[H^+] = [OH^-]$.
As both $[H^+]$ and $[OH^-]$ increases by the same amount, the value of K_w increases as well.
At 373 K, the value of $K_w = 5.13 \times 10^{-13}$ mol^2 dm^{-6}
A **neutral** solution is defined as one in which **$[H^+] = [OH^-]$**.
At **100°C**,
$[H^+] = [OH^-] = \sqrt{K_w} = \sqrt{5.13 \times 10^{-13}} = 7.16 \times 10^{-7}$ mol dm^{-3}
pH $= -\log(7.16 \times 10^{-7}) =$ **6.15**
i.e. the pH of a neutral solution is only 7.00 at 298 K so the temperature should always be stated.

9.1 Oxidation and reduction
Syllabus
Nature of science – can you relate this topic to these concepts?
How evidence is used – changes in the definition of oxidation and reduction from one involving specific elements (oxygen and hydrogen), to one involving electron transfer, to one invoking oxidation numbers is a good example of the way that scientists broaden similarities to general principles.

Understandings – how well can you explain these statements?
Oxidation and reduction can be considered in terms of oxygen gain/hydrogen loss, electron transfer or change in oxidation number.
An oxidizing agent is reduced and a reducing agent is oxidized.
Variable oxidation numbers exist for transition metals and for most main-group non-metals.
The activity series ranks metals according to the ease with which they undergo oxidation.
The Winkler Method can be used to measure biochemical oxygen demand (BOD), used as a measure of the degree of pollution in a water sample.

Applications and skills – how well can you do all of the following?
Deduction of the oxidation states of an atom in an ion or a compound.
Deduction of the name of a transition metal compound from a given formula, applying oxidation numbers represented by Roman numerals.
Identification of the species oxidized and reduced and the oxidizing and reducing agents, in redox reactions.
Deduction of redox reactions using half-equations in acidic or neutral solutions.
Deduction of the feasibility of a redox reaction from the activity series or reaction data.
Solution of a range of redox titration problems.
Application of the Winkler Method to calculate BOD.

Definition 1 (addition or removal of oxygen or hydrogen)

- Oxidation: **addition of oxygen** to a species or **removal of hydrogen** from a species.
- Reduction: **removal of oxygen** from a species or **addition of hydrogen** to a species.

Examples:

1. Extraction of iron from haematite in a blast furnace.
Fe_2O_3 (s) + 3CO (g) → 2Fe (s) + 3CO_2 (g)
iron (III) oxide + carbon monoxide → iron + carbon dioxide
Iron (III) oxide **loses oxygen** and forms iron. It is **reduced**.
Carbon monoxide **gains oxygen** and forms carbon dioxide. It is **oxidised**.

2. Chlorine gas bubbled through aqueous hydrogen sulfide.
H_2S (aq) + Cl_2 (g) → S (s) + 2HCl (aq)
hydrogen sulfide + chlorine → sulfur + hydrogen chloride
Hydrogen sulfide **loses hydrogen**. It is **oxidised**.
Chlorine **gains hydrogen**. It is **reduced**.

Definition 2 (loss or gain of electrons)

For reactions that do not involve oxygen.
- Oxidation: a species **loses** one or more **electrons**.
- Reduction: a species **gains** one or more **electrons**.
- Oxidising agent: a species that **oxidises another species** by **removing one or more electrons** – it itself is **reduced**.
- Reducing agent: a species that **reduces another species** by **giving one or more electrons** – it itself is **oxidised**.

Examples:
1. Combustion of magnesium: $Mg\ (s) + ½\ O_2\ (g) \rightarrow MgO\ (s)$
Half equations: $Mg \rightarrow Mg^{2+} + 2e^-$ and $O + 2e^- \rightarrow O^{2-}$
Magnesium **loses 2 electrons** – it is **oxidised** and acts as the **reducing agent**.
Each oxygen atom **gains 2 electrons** – it is **reduced** and acts as the **oxidising agent**.
2. Reaction of calcium with chlorine gas: $Ca\ (s) + Cl_2\ (g) \rightarrow CaCl_2\ (s)$
Half equations: $Ca \rightarrow Ca^{2+} + 2e^-$ and $Cl_2 + 2e^- \rightarrow 2Cl^-$
Calcium **loses 2 electrons** – it is **oxidised** and acts as the **reducing agent**.
Each chlorine atom **gains 1 electron** – it is **reduced** and acts as the **oxidising agent**.

Common oxidising agents

Oxidised form	Reduced form
Cl_2 (chlorine)	Cl^- (chloride ions)
Br_2 (bromine)	Br^- (bromide ions)
I_2 in KI solution (iodine), with starch	I^- (iodide ions)
Fe^{3+} (iron (III) ions)	Fe^{2+} (iron (II) ions)
H^+ (dilute acids)	H_2 (hydrogen)
H_2SO_4 (conc. sulfuric acid)	SO_2 (sulfur dioxide)
HNO_3 (conc. nitric acid)	NO_2 (nitrogen dioxide)
$Cr_2O_7^{2-}$ (dichromate (VI) ions) in H_2SO_4 (aq)	Cr^{3+} (chromium (III) ions)
MnO_4^- (manganate (VII) ions) in H_2SO_4 (aq)	Mn^{2+} (manganese (II) ions)
H_2O_2 (hydrogen peroxide) in H_2SO_4 (aq)	H_2O (water)

Common reducing agents

Reducing Agent	Product(s)
I^- (iodide ions)	I_2 (iodine)
HI (hydrogen iodide)	I_2 (iodine)
H_2S (hydrogen sulfide)	S (sulfur)
SO_2 (sulfur dioxide)	SO_4^{2-} (sulfate ions)
Fe^{2+} (iron (II) ions)	Fe^{3+} (iron (III) ions)
H_2O_2 (hydrogen peroxide) in H_2SO_4 (aq)	O_2 (oxygen)
Sn^{2+} (tin (II) ions)	Sn^{4+} (tin (IV) ions)
C (carbon)	CO (carbon monoxide) or CO_2 (carbon dioxide)
CO (carbon monoxide)	CO_2 (carbon dioxide)
Metals e.g. Mg, Zn	Metal ions e.g. Mg^{2+}, Zn^{2+}

Definition 3 (change of oxidation state)

The oxidation state of an atom in a free element, compound (molecule) or ion is the charge that the atom would have if the compound or ion were fully ionic.

Element	Oxidation state
Li, Na, K	+1
Mg, Ca, Sr, Ba	+2
Al	+3
N	-3, +3, +5
O (except in peroxides, superoxides or with F)	-2
S	-2, +1, +2, +2 ½, +4, +6
F	-1
Cl, Br, I (except with O in oxoanions and oxoacids)	-1
Cl, Br, I (with O in oxoanions and oxoacids)	+1, +3, +5, +7

Bonding electrons are assigned to the **more electronegative element** e.g.
H_2S: S is more electronegative than H, so S gains electrons (oxidation number -2) and H loses electron (oxidation number +1)
BrO_3^-: O is more electronegative than Br, so O gains electrons (oxidation number -2) and Br loses electron (oxidation number +5)

Examples:

What is the oxidation number of chlorine in $NaCl$, ClO^- and ClO_3^-?
Answer: -1, +1, +5

What is the oxidation number of sulfur in SCl_2, SO_2 and SO_4^{2-}?
Answer: +2, +4, +6

What is the oxidation number of manganese in MnO_4^- and MnO_4^{2-}?
Answer: +7, +6

What is the oxidation number of vanadium in VO_2^+?
Answer: +5

For an element, it is:

- **Oxidised** if there is an **increase** in its **oxidation number**.
- **Reduced** if there is a **decrease** in its **oxidation number**.

Examples of oxidation:

Fe^{2+} (aq) → Fe^{3+} (aq) + e^-
Fe: +2 to +3
H_2S (g) → S (s) + $2H^+$ (aq) + $2e^-$
S: -2 to 0
$2I^-$ (aq) → I_2 (s) + $2e^-$
I: -1 to 0

Examples of reduction:

MnO_4^- (aq) + $8H^+$ (aq) + $5e^-$ → Mn^{2+} (aq) + $4H_2O$ (l)
Mn: +7 to +2
PbO_2 (s) + $4H^+$ (aq) + $2e^-$ → Pb^{2+} (aq) + $2H_2O$ (l)
Pb: +4 to +2

Compound or ions which contain element that can have more than one oxidation number – numbers are assigned.

MnO_4^- : manganate (VII)
$Cr_2O_7^{2-}$: dichromate (VI)
$FeCl_3$: iron (III) chloride
$FeSO_4$: iron (II) sulfate
ClO^- : chlorate (I)
ClO_3^- : chlorate (V)

Ionic half-equations

- Balanced for <u>charge</u> and <u>number of atoms</u>.
- Must include <u>state symbols</u>.

Mg + ½ O$_2$ → MgO
Mg (s) → Mg^{2+} (s) + 2e$^-$
½ O$_2$ (g) + 2e$^-$ → O^{2-} (s)

Ca + Cl$_2$ → CaCl$_2$
Ca (s) → Ca^{2+} (s) + 2e$^-$
Cl$_2$ (s) + 2e$^-$ → 2Cl$^-$ (s)

2Br$^-$ (aq) + Cl$_2$ (g) → Br$_2$ (g) + 2Cl$^-$(aq)
2Br$^-$ (aq) → Br$_2$ (g) + 2e$^-$
Cl$_2$ (g) + 2e$^-$ → 2Cl$^-$(aq)

Oxidising agents which require an acid medium

1. Add H$^+$ to the side of the equation with oxygen to form enough H$_2$O on the other side.
2. Balance the charges by adding electrons.
3. For reduction reactions, electrons are always on the left-hand side.

Potassium manganate(VII) solution with dilute sulfuric acid.
Manganese (Mn) in MnO$_4^-$ is reduced to Mn^{2+} ions.
MnO$_4^-$ → Mn^{2+}
Add H$^+$ to the left-hand side.
MnO$_4^-$ (aq) + 8H$^+$ (aq) → Mn^{2+} (aq) + 4H$_2$O (l)
Add electrons to balance the charges.
MnO$_4^-$ (aq) + 8H$^+$ (aq) + 5e$^-$ → Mn^{2+} (aq) + 4H$_2$O (l)

Dichromate(VI) ions in acid solution.
Chromium (Cr) in Cr$_2$O$_7^{2-}$ is reduced to Cr^{3+} ions.
Cr$_2$O$_7^{2-}$ → Cr^{3+}
Balance for number of Cr.
Cr$_2$O$_7^{2-}$ → 2Cr^{3+}
Add H$^+$ to the left-hand side.
Cr$_2$O$_7^{2-}$ (aq) + 14H$^+$ (aq) → 2Cr^{3+} (aq) + 7H$_2$O (l)
Add electrons to balance the charges.
Cr$_2$O$_7^{2-}$ (aq) + 14H$^+$ (aq) + 6e$^-$ → 2Cr^{3+} (aq) + 7H$_2$O (l)

Overall redox equations from half equations

1. Multiply the half equation(s) by whole numbers to balance the number of electrons.
2. Add the two equations together, cancelling the electrons and spectator ions.

Aqueous silver nitrate and copper metal
$Ag^+ (aq) + e^- \rightarrow Ag (s)$ ... (x2)
$Cu (s) \rightarrow Cu^{2+} (aq) + 2e^-$
$2Ag^+ (aq) + 2e^- + Cu (s) \rightarrow 2Ag (s) + Cu^{2+} (aq) + 2e^-$
$2Ag^+ (aq) + Cu (s) \rightarrow 2Ag (s) + Cu^{2+} (aq)$

Acidified dichromate (VI) and Fe^{2+} solution.
$Fe^{2+} (aq) \rightarrow Fe^{3+} (aq) + e^-$ ...(x6)
$Cr_2O_7^{2-} (aq) + 14H^+ (aq) + 6e^- \rightarrow 2Cr^{3+} (aq) + 7H_2O (l)$
$Cr_2O_7^{2-} (aq) + 14H^+ (aq) + 6e^- + 6Fe^{2+} (aq) \rightarrow 2Cr^{3+} (aq) + 7H_2O (l) + 6Fe^{3+} (aq) + 6e^-$
$Cr_2O_7^{2-} (aq) + 14H^+ (aq) + 6Fe^{2+} (aq) \rightarrow 2Cr^{3+} (aq) + 7H_2O (l) + 6Fe^{3+} (aq)$

Acidified potassium manganate (VII) and Sn^{2+} solution
$Sn^{2+} (aq) \rightarrow Sn^{4+} (aq) + 2e^-$ ...(x5)
$MnO_4^- (aq) + 8H^+ (aq) + 5e^- \rightarrow Mn^{2+} (aq) + 4H_2O$ ...(x2)
$5Sn^{2+} (aq) + 2MnO_4^- (aq) + 16H^+ (aq) + 10e^- \rightarrow 2Mn^{2+} (aq) + 8H_2O + 5Sn^{4+} (aq) + 10e^-$
$5Sn^{2+} (aq) + 2MnO_4^- (aq) + 16H^+ (aq) \rightarrow 2Mn^{2+} (aq) + 8H_2O + 5Sn^{4+} (aq)$

Half equations from overall redox equations

Nitric acid and hydrogen sulfide
$H_2S (g) \rightarrow 2H^+ (aq) + S (s) + 2e^-$ (sulfur is oxidised – loses hydrogen)
$HNO_3 (aq) + H^+ (aq) + e^- \rightarrow NO_2 (g) + H_2O (l)$ (nitric acid is reduced) ... (x2) to balance number of electrons
$2HNO_3 (aq) + 2H^+ (aq) + 2e^- \rightarrow 2NO_2 (g) + 2H_2O (l)$
$2HNO_3 (aq) + H_2S (g) \rightarrow 2NO2 (g) + S (s) + 2H_2O (l)$

Redox and oxidation numbers

In any redox reaction:
- the **reducing agent** gets **oxidised**.
- the **oxidising agent** gets **reduced**.
- the **increase in the oxidation number** of the element in the reducing agent **must equal** the **decrease in oxidation number** of the element in the oxidising agent.

Example 1

Bromate ions (BrO_3^-) react with bromide ions (Br^-) in acidic solution to form bromine and water.
Oxidation state of Br in:
BrO_3^- : Br + 3 (-2) = -1, **Br = +5 (reduced)**
Br^- : **Br = -1 (oxidised)**
Br_2 : **Br = 0**
There must be 5 Br^- to every BrO_3^-. Add H^+ to balance charge and because it is in acidic solution. Add H_2O to balance the equation.
5Br^- (aq) + BrO_3^- (aq) + 6H^+ (aq) → 3Br_2 (l) + 3H_2O (l)

Example 2

Dichromate (VI) ions ($Cr_2O_7^{2-}$) react with iodide ions (I^-) in acidic solution to form iodine.
Oxidation state of Cr in:
$Cr_2O_7^{2-}$: 2Cr + 7 (-2) = -2 **Cr = +6**
Cr^{3+} : **Cr = +3 (reduced)**
Total change (2 Cr atoms per $Cr_2O_7^{2-}$ ion): **2 x +3 = +6**
Oxidation state of I in:
I^- : **I= -1**
I_2 : **I= 0 (oxidised)**
There must be 6 I^- to every $Cr_2O_7^{2-}$. Add H^+ to balance charge and because it is in acidic solution. Add H_2O to balance the equation.
$Cr_2O_7^{2-}$ + 6I^- (aq) + 14H^+ (aq) → 2Cr^{3+} (aq) + 3I_2 (s) + 7H_2O (l)

Reactivity – Redox couples

In a redox reaction, a species which gains or loses electron(s) will form a product as a result. This pair of species is known as a redox couple.

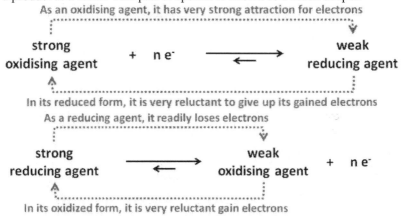

Examples of relative strengths of redox couples as oxidising and reducing agents:

	Oxidised form		Reduced form	
Increasing strength as oxidising agent	$Na^+ + e^-$	⇌	Na	Increasing strength as reducing agent
	$Mg^{2+} + 2e^-$	⇌	Mg	
	$Zn^{2+} + 2e^-$	⇌	Zn	
	$Fe^{2+} + 2e^-$	⇌	Fe	
	$2H^+ + 2e^-$	⇌	H_2	
	$Cu^{2+} + 2e^-$	⇌	Cu	
	$I_2 + 2e^-$	⇌	$2I^-$	
	$Br_2 + 2e^-$	⇌	$2Br^-$	
	$Cl_2 + 2e^-$	⇌	$2Cl^-$	
	$F_2 + 2e^-$	⇌	$2F^-$	

Strength non-metals as oxidising agents

Non-metals have a tendency to accept electrons in order to complete their valence shells as they have higher electronegativities:
- non-metals are usually strong oxidising agents (easily reduced) i.e. oxidising strength increases across the period.
- the oxidising strength decreases down the group – the atomic radius increases which causes the attraction on the incoming electrons towards the nucleus to decrease.

The oxidised form of a stronger oxidising species can oxidise (displace) the reduced form of a weaker oxidising species (or stronger reducing species) e.g.
Cl_2 (aq) + $2I^-$ (aq) → I_2 (aq) + $2Cl^-$ (aq) (√)
I_2 (aq) + $2Cl^-$ (aq) → Cl_2 (aq) + $2I^-$ (aq) (X)
F_2 (aq) + $2Br^-$ (s) → Br_2 (aq) + $2F^-$ (aq) (√)
Br_2 (aq) + $2F^-$ (aq) → F_2 (aq) + $2Br^-$ (s) (X)

Strength of metals as reducing agents

Metals have a tendency to donate electrons in order to achieve empty valence shells as they have lower electronegativities:
- metals are usually strong reducing agents (easily oxidised) i.e. reducing strength decreases across the period.
- the reducing strength increases down the group – the atomic radius increases which causes the valence electrons more easily lost as they are less attracted to the nucleus.

A more reactive metal can displace a less reactive metal from its compound e.g.
Mg (s) + Fe^{2+} (aq) → Mg^{2+} (aq) + Fe (s) (√)
Fe (s) + Mg^{2+} (aq) → Fe^{2+} (aq) + Mg (s) (X)
Zn (s) + Pb^{2+} (aq) → Zn^{2+} (aq) + Pb (s) (√)
Pb (s) + Zn^{2+} (aq) → Pb^{2+} (aq) + Zn (s) (X)

The position of a metal in the series relative to hydrogen can also be used to predict whether a metal will liberate hydrogen from dilute acids.
Metals above hydrogen are powerful enough reducing agents to reduce hydrogen ions to hydrogen e.g.
Mg (s) + 2H$^+$ (aq) → H$_2$ (g) + Mg^{2+} (aq)
2Al (s) + 6H$^+$ (aq) → 3H$_2$ (g) + 2Al^{3+} (aq)

Metals below hydrogen, such as copper, are not strong enough reducing agents to reduce hydrogen ions and hence will not react with dilute acids.

The electrochemical series for metals
potassium (K) – most reactive
sodium (Na)
lithium (Li)
barium (Ba)
strontium (Sr)
calcium (Ca)
magnesium (Mg)
aluminium (Al)
titanium (Ti)
manganese (Mn)
zinc (Zn)
iron (Fe)
cadmium (Cd)
nickel (Ni)
tin (Sn)
lead (Pb)
HYDROGEN
copper (Cu)
mercury (Hg)
silver (Ag)
gold (Au)
platinum (Pt) – least reactive

Displacement reactions - metals

Zinc foil dipped into aqueous copper sulfate:
- zinc acquires a reddish-brown coating of metallic copper.
- blue colour of the solution fades as zinc sulfate is colourless.

Copper wire dipped into aqueous silver nitrate:
- copper acquires a metallic silver coating.
- solution turns blue as copper nitrate is formed.

Displacement reactions - halogens

Halides are displaced from solution by a stronger oxidising halogen. The reactions and their equations have been outlined in Topic 3 Periodicity.
- the aqueous colour of the halogen appears.
- the colour is more pronounced when dissolved in an organic (cyclohexane) layer – iodine appears purple in the top layer.

Redox titration

The concentration of oxidising and reducing agents can be estimated by titrating them with a suitable reagent.

Redox titration with oxidising agents

Cu^{2+} oxidises I^- to form iodine (Cu^{2+} reduced to Cu^+):
$2Cu^{2+}$ (aq) + $4I^-$ (aq) → $2CuI$ (aq) + I_2 (aq)
Fe^{3+} oxidises I^- to form iodine (Fe^{3+} reduced to Fe^{2+}):
$2Fe^{3+}$ (aq) + $2I^-$ (aq) → $2Fe^{2+}$ (aq) + I_2 (aq)
The liberated iodine is then titrated against a standard thiosulfate solution:
I_2 (aq) + $2Na_2S_2O_3$ (aq) → $2NaI$ (aq) + $Na_2S_4O_6$ (aq)
2 mol Cu^{2+} ≡ 1 mol I_2 ≡ 2 mol $Na_2S_2O_3$
i.e. 1 mol Cu^{2+} ≡ 1 mol $Na_2S_2O_3$
2 mol Fe^{3+} ≡ 1 mol I_2 ≡ 2 mol $Na_2S_2O_3$
i.e. 1 mol Fe^{3+} ≡ 1 mol $Na_2S_2O_3$

Example 1:

A brass (alloy of copper and zinc) screw weighing 2.19 g is dissolved in concentrated nitric acid in a fume cupboard. This forms soluble nitrates, NO_2 (g) and H_2O (l).
The excess acid was neutralised and the solution and washings were made up to 250.00 cm³ with distilled water. 25.00 cm³ samples were added to excess solid potassium iodide and titrated with a 0.106 mol dm⁻³ sodium thiosulfate solution, adding starch as the indicator near the end-point. The mean titre was 23.75 cm³.
Zn^{2+} does not oxidise iodide ions.
$2Cu^{2+}$ (aq) + $4I^-$ (aq) → $2CuI$ (aq) + I_2 (aq)
I_2 (aq) + $2Na_2S_2O_3$ (aq) → $2NaI$ (aq) + $Na_2S_4O_6$ (aq)
moles $Na_2S_2O_3$ = 0.106 x (23.75 / 1000) = 0.0025175 mol
moles Cu^{2+} in 25.00 cm³ = 0.0025175 mol
moles Cu^{2+} in 250.00 cm³ = 0.025175 mol
mass of copper in screw = 0.025175 x 63.45 = 1.599 g
% copper in screw = (1.599 / 2.19) x 100 = 73.0 %

Example 2:

25.00 cm^3 of iron(III) chloride solution was added to excess acidified potassium iodide solution. The liberated iodine required 24.20 cm^3 of 0.100 mol dm^{-3} sodium thiosulfate solution to remove the colour.
2Fe^{3+} (aq) + 2I$^-$ (aq) → 2Fe^{2+} (aq) + I$_2$ (aq)
I$_2$ (aq) + 2Na$_2$S$_2$O$_3$ (aq) → 2NaI (aq) + Na$_2$S$_4$O$_6$ (aq)
moles Na$_2$S$_2$O$_3$ = 0.100 x (24.20 / 1000) = 0.002420 mol
moles Fe^{3+} = 0.002420 mol
concentration of FeCl$_3$ = 0.002420 / (25.00/1000) = 0.0968 mol dm^{-3}

Redox titration with reducing agents

Fe^{2+} reduces MnO$_4^-$:
5Fe^{2+} (aq) + MnO$_4^-$ (aq) + 8H$^+$ (aq) → 5Fe^{3+} (aq) + Mn^{2+} (aq) + 8H$_2$O (l)
The iron(II) solution is titrated against a standard potassium manganate(VII) solution. No indicator is required because of the intense purple colour of potassium manganate(VII). The end-point is a faint pink coloured solution.
5 mol Fe^{2+} ≡ 1 mol MnO$_4^-$

Example:

Iron tablets contain FeSO$_4$. A tablet weighing 10.31 g was crushed and dissolved in 25.00 cm^3 dilute sulfuric acid. The solution was made up to 250.00 cm^3 with distilled water. 25.00 cm^3 portions of this solution was titrated with 0.0202 mol dm^{-3} potassium manganate(VII) solution. The mean titre was 26.20 cm^3.
moles MnO$_4^-$ = 0.0202 x (26.20 / 1000) = 0.0005292 mol
moles Fe^{2+} in 25.00 cm^3 = 0.0005292 x 5 = 0.002646 mol
moles Fe^{2+} in 250.00 cm^3 = 0.02646 mol
mass of iron in 1 tablet = 0.02646 x 55.8 = 1.476 g
% iron in tablet = (1.476 / 10.31) x 100 = 14.3 %

The Winkler method to determine Biochemical Oxygen Demand (BOD)

Introduction

The solubility of oxygen in water decreases when the temperature increases. In order for water to sustain aquatic life, the concentration of oxygen should be at least 6 ppm. The main competitor to aquatic life for oxygen supply in water is bacteria. These bacteria thrive by oxidizing organic matter in the water.

The biochemical oxygen demand (BOD) is defined as the amount of oxygen required to oxidise organic matter in a sample of water at 20°C over 5 days.

In the Winkler method, the reactions can be summarised as follows:
$2Mn(OH)_2 (s) + O_2 (g) \rightarrow 2MnO(OH)_2 (s)$
$MnO(OH)_2 (s) + 4H^+ (aq) + 2I^- (aq) \rightarrow Mn^{2+} (aq) + I_2 (aq) + 3H_2O (l)$
$I_2 (aq) + 2S_2O_3^{2-} (aq) \rightarrow 2I^- (aq) + S_4O_6^{2-} (aq)$
2 mol $S_2O_3^{2-} \equiv$ 1 mol I_2
1 mol $I_2 \equiv$ 1 mol $MnO(OH)_2$
2 mol $MnO(OH)_2 \equiv$ 1 mol O_2 (g)
i.e. 4 mol $S_2O_3^{2-} \equiv$ 2 mol $I_2 \equiv$ 2 mol $MnO(OH)_2 \equiv$ 1 mol O_2 (g)

Example:

25.00 cm³ of a water sample was analysed using the Winkler method, 5.00 x 10⁻⁵ mol of sodium thiosulfate was used in the final step. Calculate the BOD for this sample.
$2Mn(OH)_2 (s) + O_2 (g) \rightarrow 2MnO(OH)_2 (s)$
$MnO(OH)_2 (s) + 4H^+ (aq) + 2I^- (aq) \rightarrow Mn^{2+} (aq) + I_2 (aq) + 3H_2O (l)$
$I_2 (aq) + 2S_2O_3^{2-} (aq) \rightarrow 2I^- (aq) + S_4O_6^{2-} (aq)$
2 mol $S_2O_3^{2-} \equiv$ 1 mol I_2
1 mol $I_2 \equiv$ 1 mol $MnO(OH)_2$
2 mol $MnO(OH)_2 \equiv$ 1 mol O_2 (g)
i.e. 4 mol $S_2O_3^{2-} \equiv$ 2 mol $I_2 \equiv$ 2 mol $MnO(OH)_2 \equiv$ 1 mol O_2 (g)
Moles $O_2 =$ ¼ x moles $S_2O_3^{2-} =$ 1.25 x 10⁻⁵ mol
Mass $O_2 =$ 1.25 x 10⁻⁵ x 32.00 = 4.00 x 10⁻⁴ g
Mass of O_2 in 1 dm³ of the water sample = ~~1000/25 x 4.00 x 10⁻⁴~~ = 0.016 g = 16 mg
BOD = 16 ppm

$$\frac{4.00 \times 10^{-4} g}{25.00 \text{ cm}^3} \times \frac{1000 \text{ cm}^3}{1 \text{ dm}^3}$$

9.2 Electrochemical cells
Syllabus
Nature of science – can you relate this topic to these concepts?
Ethical implications of research – the desire to produce energy can be driven by social needs or profit.

Understandings – how well can you explain these statements?
Voltaic (Galvanic) cells:
Voltaic cells convert energy from spontaneous, exothermic chemical processes to electrical energy.
Oxidation occurs at the anode (negative electrode) and reduction occurs at the cathode (positive electrode) in a voltaic cell.
Electrolytic cells:
Electrolytic cells convert electrical energy to chemical energy, by bringing about non-spontaneous processes.
Oxidation occurs at the anode (positive electrode) and reduction occurs at the cathode (negative electrode) in an electrolytic cell.

Applications and skills – how well can you do all of the following?
Construction and annotation of both types of electrochemical cells.
Explanation of how a redox reaction is used to produce electricity in a voltaic cell and how current is conducted in an electrolytic cell.
Distinction between electron and ion flow in both electrochemical cells.
Performance of laboratory experiments involving a typical voltaic cell using two metal/metal-ion half-cells.
Deduction of the products of the electrolysis of a molten salt.

Preliminary definitions

Electrolyte: An aqueous solution of an ionic compound or a molten liquid that conducts electricity. The electrolyte undergoes a chemical change as a result.

Half cell: A system consisting of an electrode and an electrolyte which allows the transfer of electrons to convert electrical energy to chemical energy and vice versa.

Voltaic (galvanic) cell: A cell that converts chemical energy to electrical energy. Spontaneous chemical reactions at the half cells generate electrical voltage (current).

Electrolytic cell: A cell that converts electrical energy to chemical energy. An applied voltage (current) induces non-spontaneous chemical reactions at the half cells.

Electrode: A conductor of electricity (usually a metal or graphite) that makes electrical contact with the electrolyte. The anode is where oxidation occurs. The cathode is where reduction occurs.

Salt bridge: A medium that allows the movement of ions across it to complete the circuit between two half cells. This is usually a solution of an ionic compound.

Half Cells
Zinc half cell

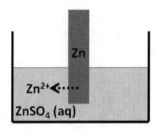

$Zn(s) \rightleftharpoons Zn^{2+}(aq) + 2e^-$

Zinc is a more reactive metal (stronger reducing agent, more easily oxidised):
- **equilibrium** lies to the **right**
- **oxidation** (loss of electrons) occurs here
  $Zn(s) \rightarrow Zn^{2+}(aq) + 2e^-$

As the zinc is oxidised, electrons are left on the zinc strip – it is negative relative to the zinc solution.

Copper half cell

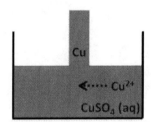

$Cu(s) \rightleftharpoons Cu^{2+}(aq) + 2e^-$

Copper is a less reactive metal (weaker reducing agent, less easily oxidised):
- **equilibrium** lies to the **left**
- **reduction** (gain of electrons) occurs here
  $Cu^{2+}(aq) + 2e^- \rightarrow Cu(s)$

As the copper ions are reduced, electrons are taken away from the copper strip – it is positive relative to the copper solution.

Remember: This happens because electrons have to flow into or out of the metal. Electrons cannot exist in the solutions!

Voltaic Cells – Half Cells combined
Daniell cell

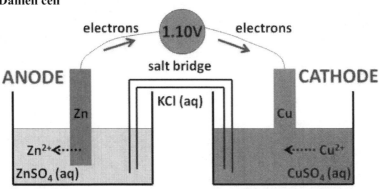

Anode is negative   Cathode is positive

As the zinc is oxidised, electrons are left on the zinc strip – it is negative relative to the zinc solution.
$Zn(s) \rightarrow Zn^{2+}(aq) + 2e^-$
As the copper ions are reduced, electrons are taken away from the copper strip – it is positive relative to the copper solution.
$Cu^{2+}(aq) + 2e^- \rightarrow Cu(s)$
These electrons flow through the external wire.
Anions (Cl⁻) move towards the anode and cations (K⁺) move towards the cathode through the salt bridge.
Overall reaction: **$Zn(s) + Cu^{2+}(aq) \rightarrow Zn^{2+}(aq) + Cu(s)$**
Salt bridge: Glass tube containing gel or absorptive paper containing a solution of ions. It completes the circuit. Na_2SO_4 or KCl is often used as it does not interfere with the reactions.
Tip:
OXIDATION at **A**NODE (**vowels**)
REDUCTION at **C**ATHODE (**consonants**)

Other voltaic cells

By combining different half cells together, voltaic cells generating different potential differences (voltages) can be created. Examples:

$Al\ (s) \rightarrow Al^{3+}\ (aq) + 3e^-$ ANODE
$Ag^+\ (aq) + e^- \rightarrow Ag\ (s)$ CATHODE
$Al\ (s) + 3Ag^+\ (aq) \rightarrow Al^{3+}\ (aq) + 3Ag\ (s)$
$Al\ (s)\ |\ 3Ag^+\ (aq)\ ||\ Al^{3+}\ (aq)\ |\ 3Ag\ (s)$
Voltage = + 2.46 V

$Zn\ (s) \rightarrow Zn^{2+}\ (aq) + 2e^-$ ANODE
$Ag^+\ (aq) + e^- \rightarrow Ag\ (s)$ CATHODE
$Zn\ (s) + 2Ag^+\ (aq) \rightarrow Zn^{2+}\ (aq) + 2Ag\ (s)$
$Zn\ (s)\ |\ 2Ag^+\ (aq)\ ||\ Zn^{2+}\ (aq)\ |\ 2Ag\ (s)$
Voltage = + 1.56 V

$Cu\ (s) \rightarrow Cu^{2+}\ (aq) + 2e^-$ ANODE
$Ag^+\ (aq) + e^- \rightarrow Ag\ (s)$ CATHODE
$Cu\ (s) + 2Ag^+\ (aq) \rightarrow Cu^{2+}\ (aq) + 2Ag\ (s)$
$Cu\ (s)\ |\ 2Ag^+\ (aq)\ ||\ Cu^{2+}\ (aq)\ |\ 2Ag\ (s)$
Voltage = + 0.46 V

In cell diagram (notation), | represents a **phase boundary** or **junction** and || represents the **salt bridge**
N.B. The further apart the metals are in the reactivity series, the larger the voltage. By convention, the anode (oxidation electrode) is always drawn on the left side in a diagram.

Electrolysis of molten sodium chloride (Downs cell)

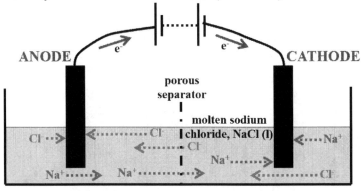

$2Cl^- (l) \rightarrow Cl_2 (g) + 2e^-$
- negative chloride ions are attracted to the positive anode.
- they lose electrons and are oxidized to chlorine gas (yellow-green gas).

$Na^+ (l) + e^- \rightarrow Na (l)$
- positive sodium ions are attracted to the negative cathode.
- they gain electrons and are reduced to sodium metal (shiny liquid).

Overall reaction: $2NaCl (l) \rightarrow 2Na (l) + Cl_2 (g)$

Electrolysis of molten lead bromide (carbon electrodes)

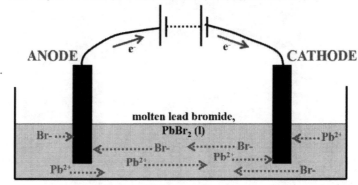

2Br- (l) → Br$_2$ (g) + 2e$^-$
- negative bromide ions are attracted to the positive anode.
- they lose electrons and are oxidized to bromine liquid (reddish-brown liquid).

Pb^{2+} (l) + 2e$^-$ → Pb (l)
- positive lead ions are attracted to the negative cathode.
- they gain electrons and are reduced to lead metal (shiny liquid).

Overall reaction: **PbBr$_2$ (l) → Pb (l) + Br$_2$ (g)**

Comparison between electrolytic and voltaic cells

	Electrolytic Cell	Voltaic Cell
Function	An applied voltage induces non-spontaneous redox reactions	Generates electrical voltage from spontaneous redox reactions at the half cells
Anode	Positive	Negative
Reaction at anode	Oxidation	Oxidation
Cathode	Negative	Positive
Reaction at cathode	Reduction	Reduction
Salt bridge	Not required – mobile ions in the electrolyte are the charge carriers	Required – to complete the circuit by flow of inert ions
External power supply	Required	Not required

19.1 Electrochemical cells
Syllabus
Nature of science – can you relate this topic to these concepts?
Employing quantitative reasoning – electrode potentials and the standard hydrogen electrode. Collaboration and ethical implications – scientists have collaborated to work on electrochemical cell technologies and have to consider the environmental and ethical implications of using fuel cells and microbial fuel cells.

Understandings – how well can you explain these statements?
A voltaic cell generates an electromotive force (EMF) resulting in the movement of electrons from the anode (negative electrode) to the cathode (positive electrode) via the external circuit. The EMF is termed the cell potential ($E^\ominus$).

The standard hydrogen electrode (SHE) consists of an inert platinum electrode in contact with 1 mol dm^{-3} hydrogen ion and hydrogen gas at 100 kPa and 298 K. The standard electrode potential ($E^\ominus$) is the potential (voltage) of the reduction half-equation under standard conditions measured relative to the SHE. Solute concentration is 1 mol dm^{-3} or 100 kPa for gases. $E^\ominus$ of the SHE is 0 V.

When aqueous solutions are electrolysed, water can be oxidized to oxygen at the anode and reduced to hydrogen at the cathode.

$\Delta G^\ominus = -nFE^\ominus$. When $E^\ominus$ is positive, $\Delta G^\ominus$ is negative indicative of a spontaneous process. When $E^\ominus$ is negative, $\Delta G^\ominus$ is positive indicative of a non-spontaneous process. When $E^\ominus$ is 0, then $\Delta G^\ominus$ is 0.

Current, duration of electrolysis and charge on the ion affect the amount of product formed at the electrodes during electrolysis.

Electroplating involves the electrolytic coating of an object with a metallic thin layer.

Applications and skills – how well can you do all of the following?
Calculation of cell potentials using standard electrode potentials.
Prediction of whether a reaction is spontaneous or not using $E^\ominus$ values.
Determination of standard free-energy changes ($\Delta G^\ominus$) using standard electrode potentials.
Explanation of the products formed during the electrolysis of aqueous solutions.
Perform lab experiments that could include single replacement reactions in aqueous solutions.
Determination of the relative amounts of products formed during electrolytic processes.
Explanation of the process of electroplating.

Preliminary definitions

Electromotive force (EMF): This is the energy supplied by a source divided by the electrical charge transported through the source (EMF = E/Q). In a voltaic cell, this is the **maximum potential (voltage)** that a cell can generate.

Cell potential ($E^{\ominus}_{cell}$): This is the potential difference between the anode and cathode of the cell. Typically, a cell rated at a certain EMF will always provide a potential difference which is **lower** due to **internal resistance** within the cell.

The potential of half cells are measured against a reference standard which is the standard hydrogen electrode (SHE).

The potential differences of electrodes relative to the SHE are known as **standard electrode potentials ($E^{\ominus}$)**.

Standard hydrogen electrode (SHE)

The potential of half cells are measured against a reference standard which is the standard hydrogen electrode (SHE). The potential differences of electrodes relative to the SHE are known as **standard electrode potentials ($E^{\ominus}$)**. The SHE is a gas electrode and consists of:

- **Hydrogen** gas at **1 atm (100 kPa)** pressure bubbling over a **platinum** plate in **1.0 mol dm^{-3} H^{+}** solution e.g. 1.0 mol dm^{-3} HCl.
- **'Platinum black'** or **'platinised platinum'** is very finely divided platinum, which **catalyses** the electrode equilibrium – the **large surface area** increases the adsorption of hydrogen gas without reacting with gas (**platinum is very inert**).

The reaction is reversible and depends on the half-cell which is connected to it.

$2H^{+} (aq) + 2e^{-} \rightleftharpoons H_2 (g) \quad E^{\ominus} = 0 \text{ V}$

$H^{+} (aq) \mid H_2 (g) \mid Pt$ or $Pt \mid H_2 (g) \mid H^{+} (aq)$

Standard electrode potential ($E^\ominus$)

This is the electrical potential of a half-cell relative to a **standard hydrogen electrode (SHE)** under these conditions:
- all solutions are at **1.0 mol dm^{-3}** concentration.
- all gases are at **1 atm (100 kPa)** pressure.
- a **stated temperature**, usually 25°C or 298 K.
- all substances are pure.
- for relatively unreactive metals, the electrode is usually the pure metal itself in a solution of 1.0 mol dm^{-3} of its salt solution.
- for non-metals, more reactive metals and a mixture of ions, a platinum electrode is used.

Measuring Standard Electrode Potential ($E^\ominus$) of metal half-cells

The half-cell is connected to a standard electrode e.g. SHE or calomel via a salt bridge.
$E^\ominus_{cell} = E^\ominus_{SHE} - E^\ominus_{half-cell}$
$E^\ominus$ (Zn^{2+}(aq)/Zn(s)) = 0 - (+0.76) = -0.76 V
For zinc, the half cell is a zinc rod dipping in a 1.0 mol dm^{-3} solution of Zn^{2+} at 298 K.
For iron (Fe^{2+}/Fe), the half cell is an iron rod dipping in a 1.0 mol dm^{-3} solution of Fe^{2+} at 298 K.

Examples:

Half-cell is positive terminal: the sign of the standard electrode potential is positive (reduction occurs) e.g. Ag^+ (aq) / Ag (s) is + 0.80 V
Half-cell (reduction): $2Ag^+$ (aq) + 2e$^-$ → 2Ag (s)
SHE (oxidation): H_2 (g) → $2H^+$ (aq) + 2e$^-$
Overall: $2Ag^+$ (aq) + H_2 (g) → 2Ag (s) + $2H^+$ (aq)
Notation: H_2 (g) | H^+ (aq) || Ag^+ (aq) | Ag (s)

Half-cell is negative terminal: the sign of the standard electrode potential is negative (oxidation occurs) e.g. Zn^{2+} (aq) / Zn (s) is – 0.76 V
Half-cell (oxidation): Zn (s) → Zn^{2+} (aq) + 2e$^-$
SHE (reduction): $2H^+$ (aq) + 2e$^-$ → H_2 (g)
Overall: Zn (s) + $2H^+$ (aq) → Zn^{2+} (aq) + H_2 (g)
Notation: Zn (s) | Zn^{2+} (aq) || H^+ (aq) | H_2 (g)

Measuring Standard Electrode Potential ($E^\ominus$) of gases

The half-cell is a platinum plate dipping into a 1.0 mol dm^{-3} solution of the appropriate ions with the gas of 100 kPa pressure bubbling through it.
For chlorine, the gas is bubbled through a 1.0 mol dm^{-3} solution of sodium chloride.

Measuring Standard Electrode Potential ($E^\ominus$) of a mixture of ions

The half-cell is a platinum plate dipping into 1.0 mol dm^{-3} solutions.

For Fe^{3+} (aq) + e^- ⇌ Fe^{2+} (aq), the half cell is a platinum electrode dipping in a 1.0 mol dm^{-3} solutions of Fe^{2+} and Fe^{3+} at 298 K.

For a MnO_4^- (aq) + 8H^+ (aq) + 5e^- ⇌ Mn^{2+} (aq) + 4H_2O (l), the half cell is a platinum electrode dipping in a 1.0 mol dm^{-3} solutions of Mn^{2+}, MnO_4^- and H^+ at 298 K.

Spontaneity of Redox reactions

The standard electrode potential $E^\ominus_{cell}$ is related to the Gibbs free energy ($\Delta G\ominus$) by the following equation:
$\Delta G^\ominus = -nFE^\ominus_{cell}$
n = number of moles of electrons transferred
F = Faraday's constant = 95 000 C mol^{-1}
$E^\ominus_{cell}$ and $\Delta G^\ominus$ can be used to predict the spontaneity of redox reactions.

	$E^\ominus_{cell}$	$\Delta G^\ominus$
Positive value	spontaneous	not spontaneous
Negative value	not spontaneous	spontaneous

Steps involved:

1. Identify the oxidising agent (the species that is reduced).
2. Identify the reducing agent (the species that is oxidised).
3. Reverse the appropriate half equation, changing the sign for $E^\ominus_{half\text{-}cell}$.
4. Calculate $E^\ominus_{cell}$ by adding the two values of $E^\ominus_{half\text{-}cell}$ together.
5. Calculate the Gibbs free energy using $\Delta G^\ominus = -nFE^\ominus_{cell}$

Example 1

Will dichromate(VI) ions oxidise chloride ions in an acidic solution?

$Cr_2O_7^{2-}$ (aq) + 6e$^-$ + 14H$^+$ (aq) $\rightleftharpoons$ 2Cr^{3+} (aq) + 7H$_2$O (l) $E^\ominus_{half\text{-}cell}$ = +1.33 V

Cl_2 (g) + 2e$^-$ $\rightleftharpoons$ 2Cl$^-$ (aq) $E^\ominus_{half\text{-}cell}$ = +1.36 V
$Cr_2O_7^{2-}$ (aq) is the oxidising agent (reduced to Cr^{3+}).
Cl$^-$ is the reducing agent (oxidised to Cl$_2$).
Reverse the equation for (Cl$_2$/Cl$^-$) and multiply it by 3 (note that the value of $E^\ominus_{half\text{-}cell}$ is **not multiplied** but the **sign is changed**):
6Cl$^-$ (aq) $\rightleftharpoons$ 3Cl$_2$ (g) + 6e$^-$ $E^\ominus_{half\text{-}cell}$ = -1.36 V

Overall: $Cr_2O_7^{2-}$ (aq) + 14H$^+$ (aq) + 6Cl$^-$ (aq) $\rightleftharpoons$ 2Cr^{3+} (aq) + 7H$_2$O (l) + 3Cl$_2$ (g)
$E^\ominus_{cell}$ = +1.33 + (-1.36) = -0.03 V
$\Delta G^\ominus$ = - 6 × 96 500 × -0.03 = 17370 J
$E^\ominus_{cell}$ < 0 and $\Delta G^\ominus$ > 0; the reaction will not occur spontaneously under standard conditions.

Example 2

Will manganate(VII) ions oxidise chloride ions in an acidic solution?

MnO_4^- (aq) + 5e$^-$ + 8H$^+$ (aq) $\rightleftharpoons$ Mn^{2+} (aq) + 4H$_2$O (l) $E^\ominus_{half-cell}$ = +1.51 V

Cl$_2$ (g) + 2e$^-$ $\rightleftharpoons$ 2Cl$^-$ (aq) $E^\ominus_{half-cell}$ = +1.36 V
MnO_4^- (aq) is the oxidising agent (reduced to Mn^{2+}).
Cl$^-$ is the reducing agent (oxidised to Cl$_2$).
Reverse the equation for (Cl$_2$/Cl$^-$) and multiply it by 5:

10Cl$^-$ (aq) $\rightleftharpoons$ 5Cl$_2$ (g) + 10e$^-$ $E^\ominus_{half-cell}$ = -1.36 V
Multiply the equation for (MnO$_4^-$/Mn^{2+}) by 2:

2MnO$_4^-$ (aq) + 10e$^-$ + 16H$^+$ (aq) $\rightleftharpoons$ 2Mn^{2+} (aq) + 8H$_2$O (l) $E^\ominus_{half-cell}$ = +1.51 V

Overall: 2MnO$_4^-$ (aq) + 16H$^+$ + 10Cl$^-$ (aq) $\rightleftharpoons$ 2Mn^{2+} (aq) + 8H$_2$O (l) + 5Cl$_2$ (g)
$E^\ominus_{cell}$ = -1.36 + (+1.51) = +0.15 V
$\Delta G^\ominus$ = - 10 × 96 500 × -0.15 = +144750 J
$E^\ominus_{cell}$ > 0 and $\Delta G^\ominus$ < 0; the reaction will occur spontaneously under standard conditions.

Example 3

Will iron(III) ions oxidise iodide ions in an acidic solution?

Fe^{3+} (aq) + e$^-$ $\rightleftharpoons$ Fe^{2+} (aq) $E^\ominus_{half-cell}$ = +0.77 V

I$_2$ (g) + 2e$^-$ $\rightleftharpoons$ 2I$^-$ (aq) $E^\ominus_{half-cell}$ = +0.54 V
Fe^{3+} (aq) is the oxidising agent (reduced to Fe^{2+}).
I$^-$ is the reducing agent (oxidised to I$_2$).
Reverse the equation for (I$_2$/I$^-$):

2I$^-$ (aq) $\rightleftharpoons$ I$_2$ (g) + 2e$^-$ $E^\ominus_{half-cell}$ = -0.54 V
Multiply the equation for (Fe^{3+}/Fe^{2+}) by 2:

2Fe^{3+} (aq) + 2e$^-$ $\rightleftharpoons$ 2Fe^{2+} (aq) $E^\ominus_{half-cell}$ = +0.77 V

Overall: 2I$^-$ (aq) + 2Fe^{3+} (aq) $\rightleftharpoons$ I$_2$ (g) + 2Fe^{2+} (aq)
$E^\ominus_{cell}$ = +0.77 + (-0.54) = +0.23 V
$\Delta G^\ominus$ = - 2 x 96500 × +0.23 = -44390 J
$E^\ominus_{cell}$ > 0 and $\Delta G^\ominus$ < 0; the reaction will occur spontaneously under standard conditions.

Example 4

Will manganate(VII) ions oxidise ethanedioate ions in an acidic solution?

MnO_4^- (aq) + 5e$^-$ + 8H$^+$ (aq) $\rightleftharpoons$ Mn^{2+} (aq) + 4H$_2$O (l) $E^\ominus_{half-cell}$ = +1.51 V

2CO$_2$ (g) + 2e$^-$ $\rightleftharpoons$ $C_2O_4^{2-}$ (aq) $E^\ominus_{half-cell}$ = -0.49 V

MnO_4^- (aq) is the oxidising agent (reduced to Mn^{2+}).
$C_2O_4^{2-}$ (aq) is the reducing agent (oxidised to CO_2).
Reverse the equation for ($CO_2/C_2O_4^{2-}$) and multiply it by 5:

5$C_2O_4^{2-}$ (aq) $\rightleftharpoons$ 10CO$_2$ (g) + 10e$^-$ $E^\ominus_{half-cell}$ = +0.49 V

Multiply the equation for (MnO_4^-/Mn^{2+}) by 2:

2MnO_4^- (aq) + 10e$^-$ + 16H$^+$ (aq) $\rightleftharpoons$ 2Mn^{2+} (aq) + 8H$_2$O (l) $E^\ominus_{half-cell}$ = +1.51 V

Overall: 5$C_2O_4^{2-}$ (aq) + 2MnO_4^- (aq) + 16H$^+$ (aq) $\rightleftharpoons$ 10CO$_2$ (g) + 2Mn^{2+} (aq) + 8H$_2$O (l)

$E^\ominus_{cell}$ = +1.51 + (+0.49) = +2.02 V

$\Delta G^\ominus$ = - 10 × 96500 × +2.02 = +1949300 J

$E^\ominus_{cell}$ > 0 and $\Delta G^\ominus$ < 0; the reaction will occur spontaneously under standard conditions.

Electrolysis of aqueous solutions

When a **molten salt** is electrolysed:
- cathode: the cations are attracted - they undergo reduction.
- anode: the anions are attracted - they undergo oxidation.

When an **aqueous solution** is electrolysed, **selective discharge** occurs where the water present can also be oxidised or reduced:
- cathode (water can be reduced to hydrogen gas)
$2H_2O\ (l) + 2e^- \rightarrow H_2\ (g) + 2OH^-\ (aq)$ $E^\ominus_{half-cell} = -0.83$ V
- anode (water can be oxidised to oxygen gas)
$2H_2O\ (l) \rightarrow O_2\ (g) + 4H^+\ (aq) + 4e^-$ $E^\ominus_{half-cell} = -1.23$ V

Cations (positive ions)	Reaction	Equation
higher than hydrogen in the electrochemical series ($E^\ominus < 0V$) i.e. cations are not easily reduced	water is reduced to hydrogen gas	$2H_2O\ (l) + 2e^- \rightarrow H_2\ (g) + 2OH^-\ (aq)$ $2H^+\ (aq) + 2e^- \rightarrow H_2\ (g)$
lower than hydrogen in the electrochemical series ($E^\ominus > 0V$)	the cations are reduced e.g. silver, copper	$Ag^+\ (aq) + e^- \rightarrow Ag\ (s)$ $Cu^{2+}\ (aq) + 2e^- \rightarrow Cu\ (s)$

Anions (negative ions)	Reaction	Equation
higher than oxygen in the electrochemical series ($E^\ominus < +0.40V$) i.e. anions are not easily oxidised	water is oxidised to oxygen gas	$2H_2O\ (l) \rightarrow O_2\ (g) + 4H^+\ (aq) + 4e^-$ $4OH^-\ (aq) \rightarrow O_2\ (g) + 2H_2O\ (l) + 4e^-$
lower than oxygen in the electrochemical series ($E^\ominus > +0.40V$)	the anions are oxidised e.g.	$2Br^-\ (aq) \rightarrow Br_2\ (l) + 2e^-$ $2I^-\ (aq) \rightarrow I_2\ (s) + 2e^-$

| | | bromide, iodide | |

If **carbon** or other **inert materials** are used as the electrodes, the possible products from the electrolysis of an aqueous solution are:

Cathode
- formation of the metal.
- formation of hydrogen.

The product with the most positive $E^\ominus$ will be formed.

Anode
- formation of the non-metal.
- formation of oxygen.
- oxidation of the electrode.

The product with the most negative $E^\ominus$ will be formed.

Generally:
Cathode - hydrogen is formed unless Cu^{2+} or Ag^+ are present.
Anode - oxygen is formed unless Br^- or I^- are present.

Examples of electrolysis of aqueous solutions with carbon (graphite) electrodes and their products:

Solution	Product at cathode	Product at anode
copper(II) bromide $CuBr_2$ (aq)	copper metal Cu (s)	bromine liquid Br_2 (l)
sodium iodide NaI (aq)	hydrogen gas H_2 (g)	iodine solid I_2 (s)
silver nitrate $AgNO_3$ (aq)	silver metal Ag (s)	oxygen gas O_2 (g)
potassium sulfate K_2SO_4 (aq)	hydrogen gas H_2 (g)	oxygen gas O_2 (g)
copper sulfate $CuSO_4$ (aq)	copper metal Cu (s)	oxygen gas O_2 (g)

Electrolysis of aqueous solutions (ion selection)

When the tendency to be discharge of the ion and water is similar, the **concentration of the electrolyte** may be the determining factor as to which ion is discharged at the electrode.

Example: Electrolysis of sodium chloride solution
Oxidation of chloride ions ($E^\ominus = +1.36$ V)
Oxidation of water ($E^\ominus = +1.23$ V)
Dilute aqueous sodium chloride - the product at the anode is oxygen.
Saturated solution of sodium chloride - the major product at the anode is chlorine.

Electrolysis of water using platinum electrodes

Pure water cannot of course be electrolysed - the concentration of ions is too low to allow it to conduct the current.
Dilute solutions of electrolytes producing ions that are the same as those of water e.g. H_2SO_4, NaOH, K_2SO_4 can be electrolysed.
Reaction at cathode (negative electrode):
$2H^+$ (aq) + $2e^-$ → H_2 (g)
Bubbles of colourless hydrogen gas is formed. The gas can be tested by igniting it with a lighted splint which gives a squeaky pop. The pH of the solution surround the cathode increases due to the gradual depletion of H^+ ions.
Reaction at anode (positive electrode):
$2H_2O$ (l) → O_2 (g) + $4H^+$ (aq) + $4e^-$
Bubbles of colourless oxygen gas is formed. The gas can be tested with a glowing splint which relights instantly. The pH of the solution surround the cathode decreases due to the gradual formation of H^+ ions.
Overall reaction: $2H_2O$ (l) → $2H_2$ (g) + O_2 (g)
- the production of one mole of oxygen gas releases 4 electrons
- 4 moles of electrons produces two moles of H_2 gas.

Electrolysis of concentrated brine (sodium chloride)

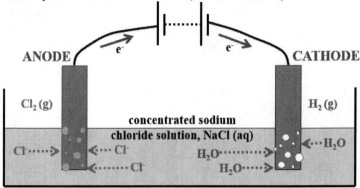

Cathode (hydrogen gas is formed as $E^\ominus_{\text{half-cell}}$ is much more positive):
$Na^+ (aq) + e^- \rightarrow Na (s)$ $E^\ominus_{\text{half-cell}} = -2.71$ V
$2H_2O (l) + 2e^- \rightarrow H_2 (g) + 2OH^- (aq)$ $E^\ominus_{\text{half-cell}} = -0.83$ V

Anode (chlorine gas is formed as $E^\ominus_{\text{half-cell}}$ are similar and $[Cl^-]$ is much higher):
$2Cl^- (aq) \rightarrow Cl_2 (g) + 2e^-$ $E^\ominus_{\text{half-cell}} = -1.36$ V
$2H_2O (l) \rightarrow O_2 (g) + 4H^+ (aq) + 4e^-$ $E^\ominus_{\text{half-cell}} = -1.23$ V

Reaction at cathode (negative electrode): $2H_2O (l) + 2e^- \rightarrow H_2 (g) + 2OH^- (aq)$ Observations: Bubbles of colourless hydrogen gas is formed. The gas can be tested by igniting it with a lighted splint which gives a squeaky pop.	Uses of **hydrogen**: Feedstock for the Haber process to produce ammonia: $N_2 (g) + 3H_2 (g) \rightleftharpoons 2NH_3 (g)$
Reaction at anode (positive electrode): $2Cl^- (aq) \rightarrow Cl_2 (g) + 2e^-$ Observations: Bubbles of pale green chlorine gas is formed. The gas can be tested with damp blue litmus paper which turns red and then white (bleached).	Uses of **chlorine**: Feedstock for producing polyvinyl chloride (PVC). Bleach. Purification of water.
Reaction in the electrolyte: $2H_2O (l) + 2Cl^- (aq) \rightarrow Cl_2 (g) + H_2 (g) + 2OH^- (aq)$ As there is a gradual depletion of Cl^- ions and increase in OH^- ions, the electrolyte will eventually have a high concentration of **sodium hydroxide**, NaOH. This is an important chemical in the manufacture of soap and paper. The pH of the electrolyte also increases.	

Electrolysis of copper(II) sulfate solution using graphite electrodes

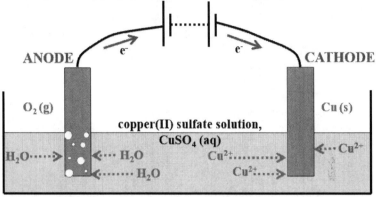

Cathode (copper metal is formed as $E^{\ominus}_{\text{half-cell}}$ is much more positive):
Cu^{2+} (aq) + 2e$^-$ → Cu (s) $E^{\ominus}_{\text{half-cell}}$ = +0.34 V
$2H_2O$ (l) + 2e$^-$ → H_2 (g) + 2OH$^-$ (aq) $E^{\ominus}_{\text{half-cell}}$ = -0.83 V
Anode (oxygen gas is formed):
$2H_2O$ (l) → O_2 (g) + 4H$^+$ (aq) + 4e$^-$ $E^{\ominus}_{\text{half-cell}}$ = -1.23 V

Reaction at cathode (negative electrode): Cu^{2+} (aq) + 2e$^-$ → Cu (s) Observations: Shiny brown layer of copper metal is formed on the surface of the electrode.
Reaction at anode (positive electrode): $2H_2O$ (l) → O_2 (g) + 4H$^+$ (aq) + 4e$^-$ Observations: Bubbles of colourless oxygen gas is formed. The gas can be tested with a glowing splint which relights instantly.
Reaction in the electrolyte: $2Cu^{2+}$ (aq) + $2H_2O$ (l) → 2Cu (s) + O_2 (g) + 4H$^+$ (aq) As there is a gradual depletion of Cu^{2+} ions and increase in H$^+$ ions, the blue colour of the electrolyte will gradually fade and the pH will decrease.

Electrolysis of copper(II) sulfate solution with copper electrodes

If the anode is made from a metal that is not inert but unreactive ($E^\ominus_{\text{half-cell}}$ > SHE) e.g. copper, the anode itself may be oxidised and dissolve into the solution. This is known as a **participant electrode**.
$E^\ominus_{\text{half-cell}}$ (Cu^{2+}/Cu) = +0.34 V is higher than $E^\ominus_{\text{half-cell}}$ (H^+/H_2) = +0.00V

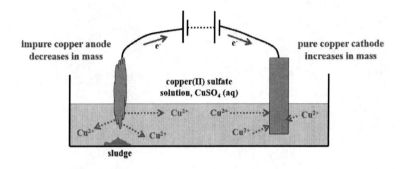

Reaction at cathode (negative electrode): Cu^{2+} (aq) + 2e⁻ → Cu (s)   Observations: Shiny brown layer of copper metal is formed on the surface of the electrode. The cathode increases in mass over time. This process is used to purify copper (electrorefining) for use as an electrical conductor.
Reaction at anode (positive electrode):   Cu (s) → Cu^{2+} (aq) + 2e⁻   Observations: The anode decreases in mass over time. At the bottom of the anode, a sludge of impurities is formed which can sometimes contain precious metals.
Reaction in the electrolyte: The equilibrium $2Cu^{2+} + 2e^- \rightleftharpoons 2Cu$ (s) is maintained. The concentration of Cu^{2+} ions does not change so there will be no observable change in the electrolyte – the intensity of the blue colour stays the same.

Electrolysis of sodium dicyanoargentate(I) solution – silver plating

Electroplating is used to coat a conducting object with a thin layer of metal. Uses include:
- decorative: silver-plated cutlery and jewellery.
- sacrificial protection: galvanized iron (iron coated in zinc) – zinc is oxidised first.
- strengthening: chromium and vanadium plating e.g. car bumpers and heavy-duty tools.

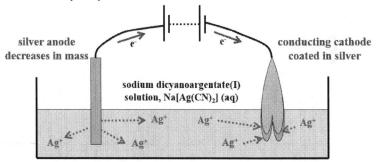

Reaction at cathode (negative electrode): $[Ag(CN)_2]^-$ (aq) + e$^-$ → Ag (s) + 2CN$^-$ (aq)   Observations: Shiny silver layer gradually forms on the surface of the conducting object. Silver nitrate, $AgNO_3$ is not used as the rate of reaction is too fast which causes poor adhesion of the silver metal.
Reaction at anode (positive electrode):   Ag (s) + 2CN$^-$ (aq) → $[Ag(CN)_2]^-$ (aq) + e$^-$ Observations: The anode decreases in mass over time. The anode does not have to be a pure metal.

Quantitative electrolysis

The amount of product that results from electrolysis will depend upon:
1. the number of electrons (n) required to produce one mole of product e.g.
   The same current is passed for the same time through aqueous solutions of Ag^+ ions and Cu^{2+} ions:
   Ag^+ (aq) + e^- → Ag (s) produces twice the amount (no. of moles) of metal at the cathode than Cu^{2+} (aq) + $2e^-$ → Cu(s)
   - the reduction of a Ag^+ ion requires one electron.
   - the reduction of a Cu^{2+} ion requires two electrons.
2. the magnitude of the current (I) i.e. the rate of flow of electrons.
3. the time (t) for which the current is passed.

To produce the same amount of copper, the number of electrons transferred must be double by either doubling the:
- current.
- time.

SI units

Quantity name	Quantity symbol	Unit (symbol)
Current	I	ampere (A)
Charge	Q	coloumb (C)
Potential difference or Voltage	V	volt (V)
Electromotive force	EMF	volt (V)
Energy	E	joule (J)

Equations

charge = current x time in seconds
Q = I x t
energy = charge x potential difference
E = Q x V
1 electron carries 1.602×10^{-19} C of negative charge.
1 mol of electrons carries a charge of
$1.602 \times 10^{-19} \times 6.023 \times 10^{23}$ = 96 488 C mol^{-1} of negative charge
This is usually approximated to 96 500 C mol^{-1} and is known as the
Faraday constant.
Knowing the charge passed and this constant, the number of moles of electrons may be calculated:
Amount of electrons (in mol) = Charge passed (in C) / 96 500
The amount of product formed can then be calculated from the balanced equation for the reaction at the electrode.

Example:

Find the mass of copper produced at the cathode by passing a current of 4.50A through aqueous copper(II) sulfate for exactly 6 hours.
Q = I x t = $4.50 \times (6 \times 60 \times 60)$ = 97 200 C
Amount of electrons = Charge passed / 96 500 = 97 200 / 96 500 = 1.0073 mol
Cu^{2+} (aq) + 2e$^-$ → Cu (s)
Amount of Cu = ½ amount of electrons = ½ × 1.0073 = 0.5036 mol
Mass of Cu = n × M_r = 0.5036 × 63.55 = 32.01 g

If the electrolysis is carried out with mains electricity with a voltage of 220V, how much energy is transferred?

E = Q x V = 97 200 x 220 = 21 384 000 J = 21.38 MJ

10.1 Fundamentals of organic chemistry

Syllabus

Nature of science – can you relate this topic to these concepts?
Serendipity and scientific discoveries – PTFE and superglue.
Ethical implications – drugs, additives and pesticides can have harmful effects on both people and the environment.

Understandings – how well can you explain these statements?
A homologous series is a series of compounds of the same family, with the same general formula, which differ from each other by a common structural unit.
Structural formulas can be represented in full and condensed format.
Structural isomers are compounds with the same molecular formula but different arrangements of atoms.
Functional groups are the reactive parts of molecules.
Saturated compounds contain single bonds only and unsaturated compounds contain double or triple bonds.
Benzene is an aromatic, unsaturated hydrocarbon.

Applications and skills – how well can you do all of the following?
Explanation of the trends in boiling points of members of a homologous series.
Distinction between empirical, molecular and structural formulas.
Identification of different classes: alkanes, alkenes, alkynes, halogenoalkanes, alcohols, ethers, aldehydes, ketones, esters, carboxylic acids, amines, amides, nitriles and arenes.
Identification of typical functional groups in molecules e.g. phenyl, hydroxyl, carbonyl, carboxyl, carboxamide, aldehyde, ester, ether, amine, nitrile, alkyl, alkenyl and alkynyl.
Construction of 3-D models (real or virtual) of organic molecules.
Application of IUPAC rules in the nomenclature of straight-chain and branched-chain isomers.
Identification of primary, secondary and tertiary carbon atoms in halogenoalkanes and alcohols and primary, secondary and tertiary nitrogen atoms in amines.
Discussion of the structure of benzene using physical and chemical evidence.

Introduction

Organic chemistry is the study of compounds based on carbon and hydrogen. Many organic compounds contain other elements such as oxygen, halogen or nitrogen.

The carbon atom (electronic configuration $1s^2 2s^2 2p^2$)

Carbon has four unpaired electrons and forms four covalent bonds which can be:
1. Four single bonds

2. Two single bonds + one double bond

$$>\!C=$$

3. One single bond + one triple bond

$$-C\equiv$$

4. Two double bonds

$$=C=$$

5. Three single bonds + one ionic bond (carbocation)

Other common elements in organic chemistry and number of bonds formed
oxygen: two bonds
hydrogen: one bond
halogens (F, Cl, Br, I): one bond

Bond geometry

Four single bonds
The four pairs of bonding electrons repel each other.

The molecule is tetrahedral with bond angles of 109.5°.

Two single bonds + one double bond
The electron pairs in the double bond occur along the same axis.
The electrons in the double bond repel the bonded electrons in the two C-H bonds e.g. ethene $H_2C=CH_2$

The molecule is planar with bond angles of 120°.
If the two CH_2 groups were rotated, one of the double bonds would break and only reform after a rotation of 180°. This does not happen unless the substance is heated very strongly. At room temperature, rotation about a double bond is not possible.

Homologous series

A **homologous series** is a series of compounds that:
- have a common **general formula**.
- differ by CH_2.
- show a trend in physical properties e.g. boiling point.
- show similar chemical properties as they have the same **functional group**.

Alkanes (general formula: C_nH_{2n+2})

Name	n	Boiling point / °C	State at r. t.
methane	1	-161	gas
ethane	2	-89	gas
propane	3	-42	gas
butane	4	-0.5	gas
pentane	5	36	liquid
hexane	6	69	liquid

Molecular formula	Condensed structural formulae
CH_4	CH_4
C_2H_6	CH_3CH_3
C_3H_8	$CH_3CH_2CH_3$
C_4H_{10}	$CH_3CH_2CH_2CH_3$
C_5H_{12}	$CH_3CH_2CH_2CH_2CH_3$
C_6H_{14}	$CH_3CH_2CH_2CH_2CH_2CH_3$

Structural formulae

$$\begin{array}{c} H \\ | \\ H-C-H \\ | \\ H \end{array}$$

$$\begin{array}{cc} H & H \\ | & | \\ H-C-C-H \\ | & | \\ H & H \end{array}$$

$$\begin{array}{ccc} H & H & H \\ | & | & | \\ H-C-C-C-H \\ | & | & | \\ H & H & H \end{array}$$

$$\begin{array}{cccc} H & H & H & H \\ | & | & | & | \\ H-C-C-C-C-H \\ | & | & | & | \\ H & H & H & H \end{array}$$

$$\begin{array}{ccccc} H & H & H & H & H \\ | & | & | & | & | \\ H-C-C-C-C-C-H \\ | & | & | & | & | \\ H & H & H & H & H \end{array}$$

$$\begin{array}{cccccc} H & H & H & H & H & H \\ | & | & | & | & | & | \\ H-C-C-C-C-C-C-H \\ | & | & | & | & | & | \\ H & H & H & H & H & H \end{array}$$

Alkenes (general formula: C_nH_{2n})

Name	n	Boiling point / °C	State at r. t.
ethene	2	-104	gas
propene	3	-48	gas
but-1-ene	4	-6	gas
pent-1-ene	5	30	liquid
hex-1-ene	6	63	liquid

Molecular formula	Condensed structural formulae
C_2H_4	CH_2CH_2
C_3H_6	CH_3CHCH_2
C_4H_8	$CH_3CH_2CHCH_2$
C_5H_{10}	$CH_3CH_2CH_2CHCH_2$
C_6H_{12}	$CH_3CH_2CH_2CH_2CHCH_2$

Structural formulae

$$\begin{array}{c} H \\ | \\ H-C=C-H \\ | \\ H \end{array}$$

$$CH_2=CH-CH_3$$

$$CH_3-CH_2-CH=CH_2$$

$$CH_3-CH_2-CH_2-CH=CH_2$$

$$CH_3-CH_2-CH_2-CH_2-CH=CH_2$$

Alcohols (general formula: $C_nH_{2n+1}OH$)

Name	n	Boiling point / °C	State at r. t.
methanol	1	65	liquid
ethanol	2	78	liquid
propan-1-ol	3	97	liquid
butan-1-ol	4	118	liquid
pentan-1-ol	5	138	liquid

Molecular formula	Condensed structural formulae
CH_3OH	CH_3OH
C_2H_5OH	CH_3CH_2OH
C_3H_7OH	$CH_3CH_2CH_2OH$
C_4H_9OH	$CH_3CH_2CH_2CH_2OH$
$C_5H_{11}OH$	$CH_3CH_2CH_2CH_2CH_2OH$

Structural formulae

$$CH_3-OH$$

$$CH_3-CH_2-OH$$

$$CH_3-CH_2-CH_2-OH$$

$$CH_3-CH_2-CH_2-CH_2-OH$$

$$CH_3-CH_2-CH_2-CH_2-CH_2-OH$$

Aldehydes (general formula: $C_nH_{2n}O$)

Name	n	Boiling point / °C	State at r. t.
methanal	1	-19	gas
ethanal	2	20	gas
propanal	3	48	liquid
butanal	4	75	liquid

Molecular formula	Condensed structural formulae
CH_2O	CH_2O
CH_3CHO	CH_3CHO
C_2H_5CHO	CH_3CH_2CHO
C_3H_7CHO	$CH_3CH_2CH_2CHO$

Structural formulae

Ketones (general formula: $C_nH_{2n}O$)

Name	n	Boiling point / °C	State at r. t.
propanone	3	56	liquid
butanone	4	80	liquid
pentan-2-one	5	102	liquid
pentan-3-one	5	102	liquid

Molecular formula	Condensed structural formulae
CH_3COCH_3	CH_3COCH_3
$C_2H_5COCH_3$	$CH_3CH_2COCH_3$
$C_3H_7COCH_3$	$CH_3CH_2CH_2COCH_3$
$C_2H_5COC_2H_5$	$CH_3CH_2COCH_2CH_3$

Structural formulae

Skeletal formulae

Full structural formula	Skeletal formula

Functional groups

A functional group is a small group of atoms or a single halogen atom that gives the compounds in a homologous series particular chemical properties.

Class	Functional group	Suffix
alkanes		-ane
alkenes	alkenyl $\diagdown_{C}=C\diagup$	-ene
alkynes	alkynyl $-C\equiv C-$	-yne
alcohols (primary)	hydroxyl —OH	-ol
alcohols (secondary)	hydroxyl —OH	-ol
alcohols (tertiary)	hydroxyl —OH	-ol

Class	General formula	First member
alkanes	C_nH_{2n+2}	methane H–CH₃ (H—C—H with H above and below)
alkenes	C_nH_{2n}	ethene H₂C=CH₂
alkynes	C_nH_{2n-2}	ethyne H—C≡C—H
alcohols (primary)	RCH_2OH	methanol H—C(H)(H)—OH
alcohols (secondary)	$RR'CHOH$	propan-2-ol H_3C—C(CH₃)(H)—OH
alcohols (tertiary)	$RR'R''COH$	2-methylpropan-2-ol H_3C—C(CH₃)(CH₃)—OH

Class	Functional group	Suffix
halogenoalkanes	halide —X (X = F, Cl, Br, I)	
aldehydes	aldehyde —C(=O)H	-al
ketones	carbonyl C=O	-one
carboxylic acids	carboxyl —C(=O)OH	-oic acid
esters	ester —C(=O)O—	-oate
ethers	ether —O—	

Class	General formula	First member
halogenoalkanes	$C_nH_{2n+1}X$	chloromethane H–C(H)(H)–Cl
aldehydes	RCHO	methanol H–CHO
ketones	RCOR'	propanone CH$_3$–CO–CH$_3$
carboxylic acids	RCOOH	methanoic acid H–COOH
esters	RCOOR'	methyl methanoate H–COO–CH$_3$
ethers	ROR'	methoxymethane H$_3$C–O–CH$_3$

Class	Functional group	Suffix
amines (primary)	amino $-N\begin{array}{c}H\\ \\H\end{array}$	-amine
amines (secondary)	amino $-N\begin{array}{c}H\\ \\ \end{array}$	-amine
amines (tertiary)	amino $-N\begin{array}{c}\\ \\ \end{array}$	-amine
amides	amido $-C\begin{array}{c}O\\ \| \\NH_2\end{array}$	-amide
nitriles	cyano $-C\equiv N$	-nitrile
arenes	phenyl $-\bigcirc$	

Class	General formula	First member
amines (primary)	RNH$_2$	methanamine H$_3$C—NH$_2$
amines (secondary)	RNHR'	N-methylmethanamine H$_3$C—NH—CH$_3$
amines (tertiary)	RNR'R''	N,N-dimethylmethanamine H$_3$C—N(CH$_3$)CH$_3$
amides	RCONH$_2$	methanamide H—C(=O)—NH$_2$
nitriles	RCN	methane nitrile (hydrogen cyanide) H—C≡N
arenes	C$_n$H$_{2n-6}$	benzene

Nomenclature

Identify the longest carbon chain – this provides the stem name. For cyclic hydrocarbons, the most complex system is the principal group.

Number of C atoms in longest chain	Stem name
1	meth-
2	eth-
3	prop-
4	but-
5	pent-
6	hex-
7	hept-
8	oct-
9	non-
10	dec-

Examples:

Longest carbon chain has six carbon atoms – the stem name is hexane.

Cyclohexane is the principal group.

Isomerism

Isomers are compounds with the same molecular formula but different structural formulae (different arrangement of atoms).

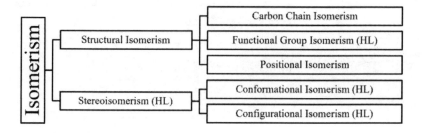

Carbon Chain Isomerism

In carbon chain isomerism, the length of carbon chain varies for the same molecular formula.

C_4H_{10} has two carbon chain isomers:

$$H_3C-CH_2-CH_2-CH_3$$
n-butane

$$H_3C-\underset{\underset{CH_3}{|}}{CH}-CH_3$$
methylpropane

C_5H_{12} has three carbon chain isomers:

$$H_3C-CH_2-CH_2-CH_2-CH_3$$
n-pentane

$$H_3C-\underset{\underset{CH_3}{|}}{CH}-CH_2-CH_3$$
2-methylbutane

$$H_3C-\underset{\underset{CH_3}{|}}{\overset{\overset{CH_3}{|}}{C}}-CH_3$$
2,2-dimethylpropane

These compounds behave the same chemically but have different physical properties.

Isomer	Boiling point / °C
n-pentane	35.9
2-methylbutane	27.8
2,2-dimethylpropane	9.0

Positional Isomerism

In positional isomerism, the position of the functional group varies between isomers.

C_4H_8 and C_5H_{10} each have two positional isomers:

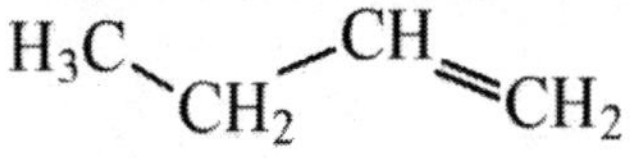

$H_3C-CH_2-CH=CH_2$
but-1-ene

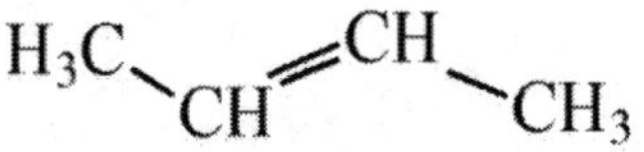

$H_3C-CH=CH-CH_3$
but-2-ene

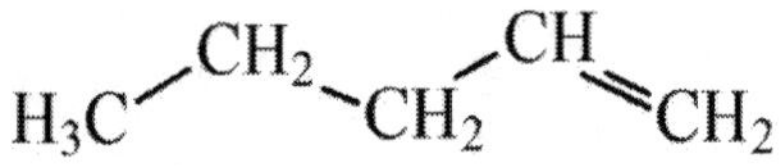

$H_3C-CH_2-CH_2-CH=CH_2$
pent-1-ene

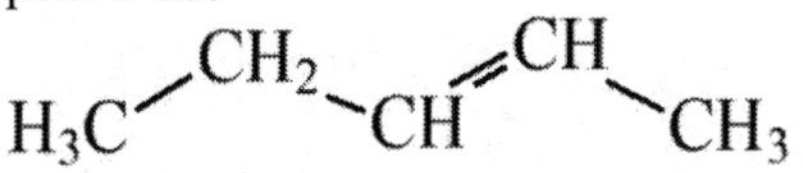

$H_3C-CH_2-CH=CH-CH_3$
pent-2-ene

The position of the functional group can influence the physical properties and chemical activity of the functional group.
- **primary (1º)** molecules: the functional group is attached to a **primary carbon atom** (carbon atom attached to hydrogen atoms only or just one alkyl group).
- **secondary (2º)** molecules: the functional group is attached to a **secondary carbon atom** (carbon atom attached to two alkyl groups).
- **tertiary (3º)** molecules: the functional group is attached to a **tertiary carbon atom** (carbon atom attached to three alkyl groups).

$$H_3C\diagup^{CH_2}\diagdown_{CH_2}\diagup^{CH_2}\diagdown OH$$

butan-1-ol (primary alcohol)

$$H_3C\diagup^{CH_2}\diagdown_{CH}\diagup^{CH_3}$$
$$\underset{HO}{|}$$

butan-2-ol (secondary alcohol)

$$H_3C-\underset{\underset{CH_3}{|}}{\overset{\overset{CH_3}{|}}{C}}-OH$$

2-methylpropan-2-ol (tertiary alcohol)

Substituents hydrocarbon chains are given the suffix –yl, designated with the smallest possible number on the longest chain and listed in alphabetical order.

Number of C atoms in substituent chain	1	2	3	4	5
Name	methyl	ethyl	propyl	butyl	pentyl

Multiple substituents of the same length are grouped as di-, tri-, tetra-, penta-, etc.

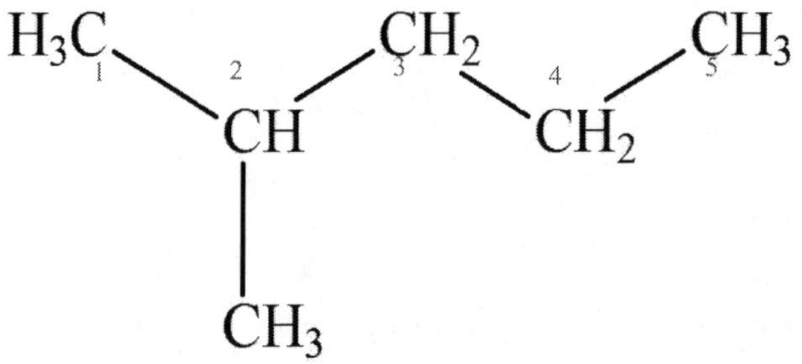

$H_3C-CH(CH_3)-CH_2-CH_2-CH_3$

2-methylpentane

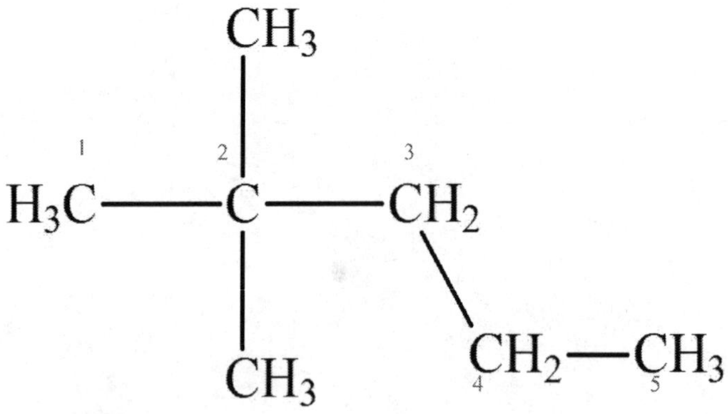

$H_3C-C(CH_3)_2-CH_2-CH_2-CH_3$

2,2-dimethylpentane

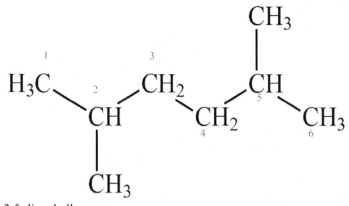

2,5-dimethylhexane

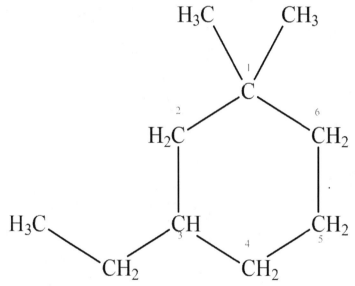

3-ethyl-1,1-dimethylcyclohexane

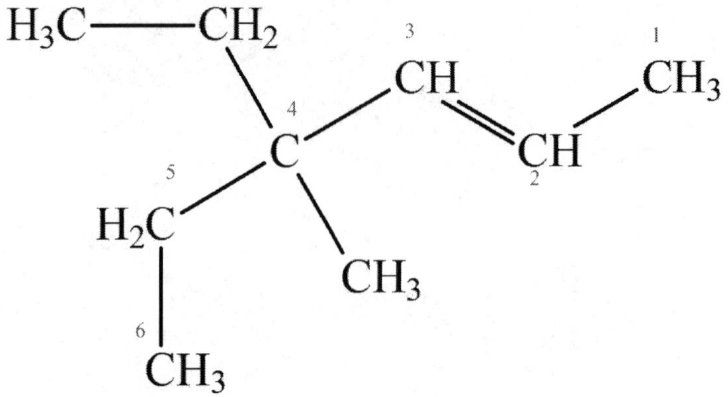

4-ethyl-4-methylhex-2-ene

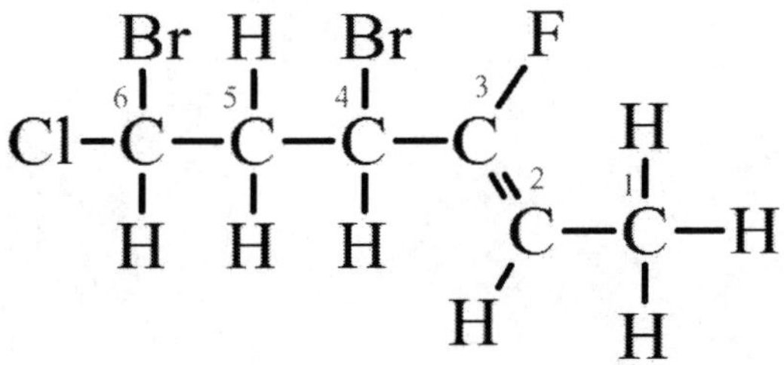

4,6-dibromo-6-chloro-3-fluorohex-2-ene

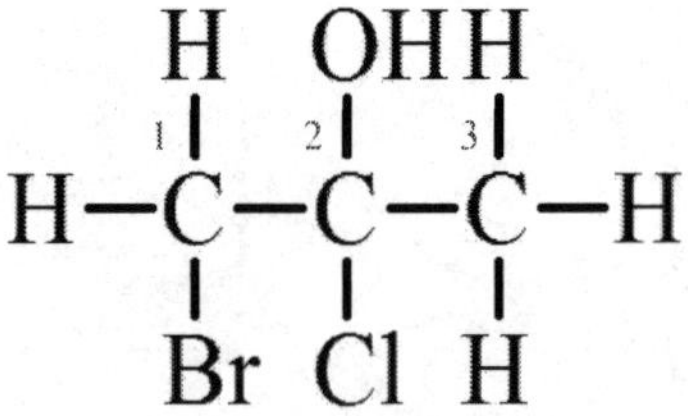

1-bromo-2-chloropropan-2-ol

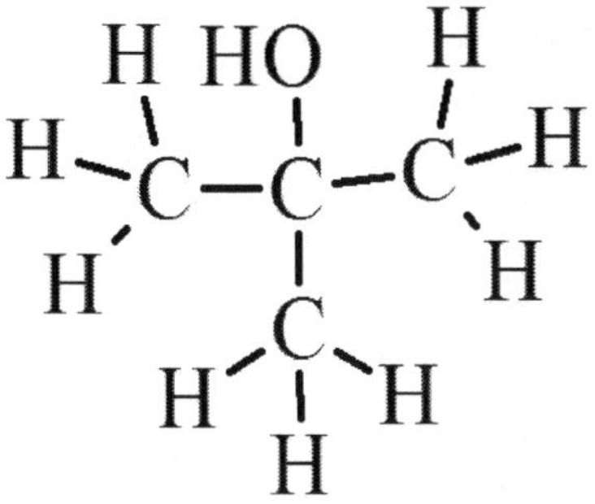

2-methylpropan-2-ol

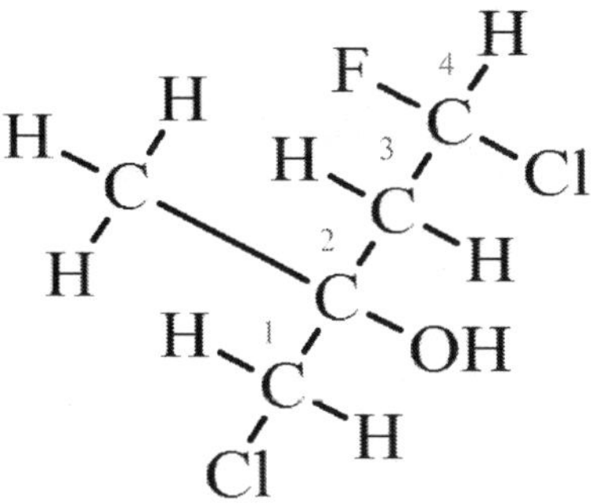

1,4-dichloro-4-fluoro-2-methylbutan-2-ol

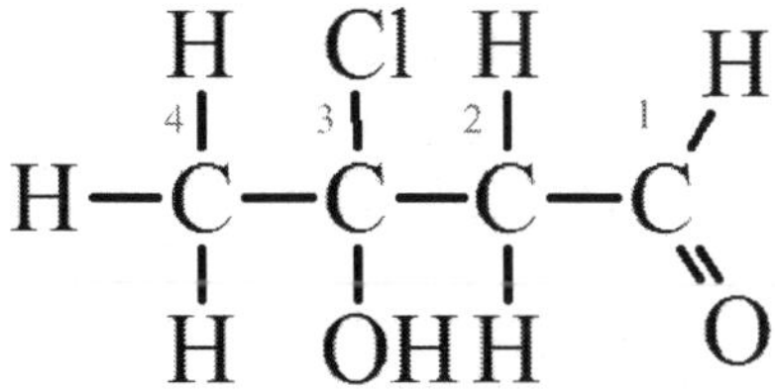

3-chloro-3-hydroxybutanal

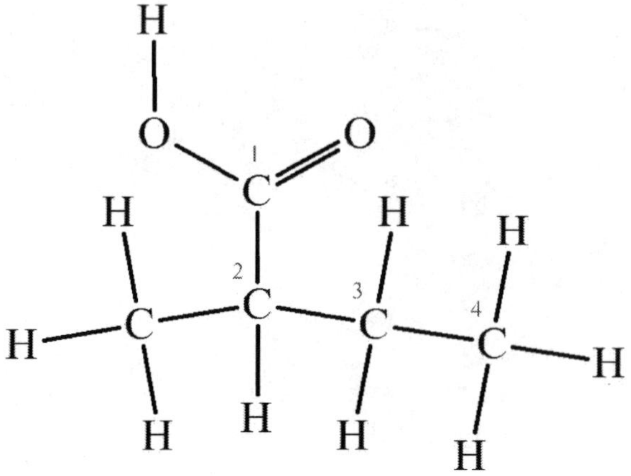

2-methylbutanoic acid

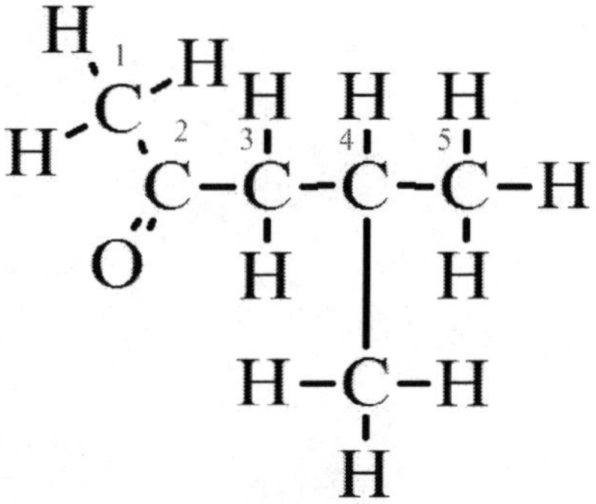

4-methylpentan-2-one

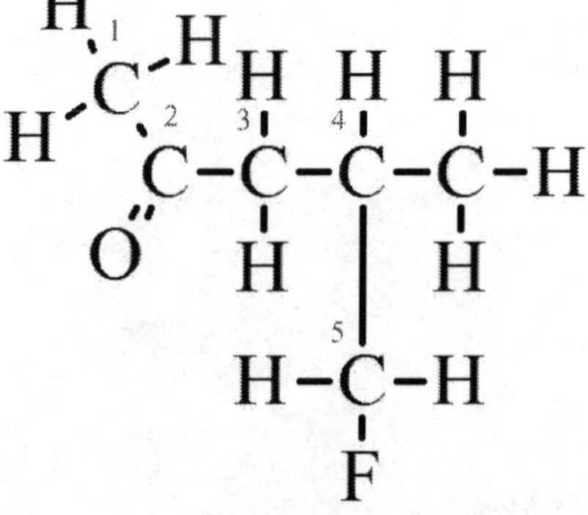

5-fluoro-4-methylpentan-2-one

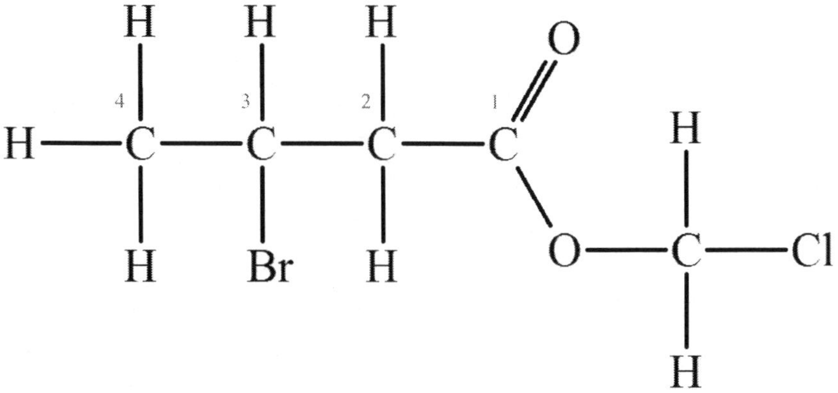

chloromethyl 3-bromobutanoate

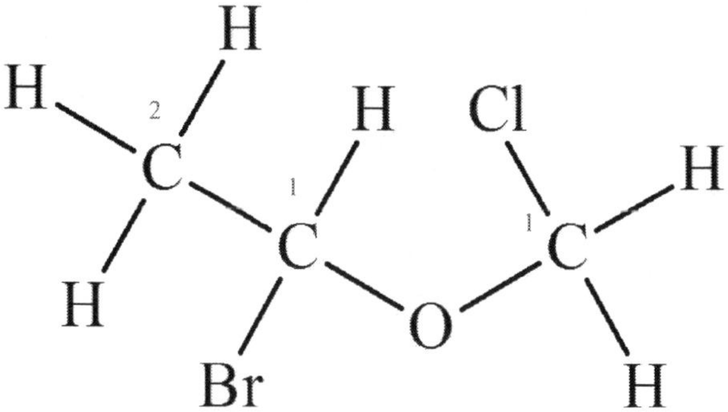

1-bromo-1-chloromethoxyethane

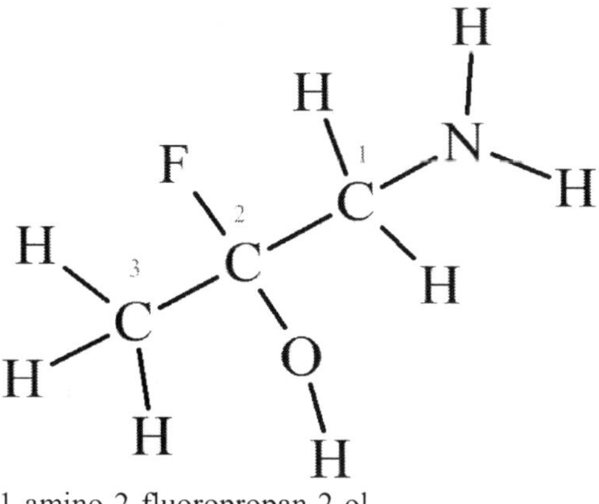

1-amino-2-fluoropropan-2-ol

N-chloromethyl-2-fluoro-N-1-fluoroethylpropan-1-amine

2-chloro-N-methylethan-1-amine

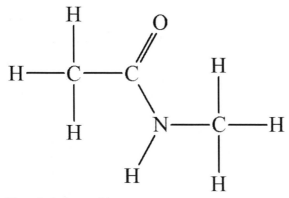
N-methylethanamide

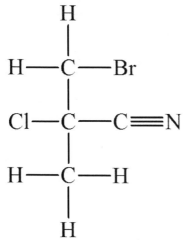
3-bromo-2-chloro-2-methylpropanenitrile

N-bromofluoromethyl-N-1,2-dichloroethylethanamine

4-bromo-2-chlorophenol

2-fluoro-4-propan-2-ylphenylamine

Aromatic hydrocarbons

Michael Faraday isolated a new hydrocarbon in 1825. It was found to have the molecular formula C_6H_6 which suggests a large number of double bonds.

In 1865, after a dream about a snake biting its own tale, Kekulé suggested the following structure for benzene which consists of alternating single and double bonds.

```
       H
   H   |   H
    \  |  /
     [ring]
    /  |  \
   H   |   H
       H
```

The structure of benzene was thought to be a resonance hybrid between two structures:

The other theory assumes that each carbon atom, which is sp² hybridized is joined by a σ-bond to each of its two neighbours, and by a third σ-bond to a hydrogen atom.
The fourth bonding electron is in a p-orbital and the six p_z-orbitals overlap above and below the ring, producing a delocalised π-system of electrons.

p_z-orbitals above and below the plane

electron density map of a benzene molecule

Aromatic hydrocarbons – evidence of benzene structure

Benzene is now commonly represented by

X-ray diffraction (bond length)
The position of the centre of the atoms show that the length of the C-C bonds in benzene are equal.

Bond	Bond length/nm
C-C bond in benzene	0.14
C-C bond in cyclohexene	0.15
C=C bond in cyclohexene	0.13

Resonance (delocalization) energy

cyclohexene + H_2 (g) → cyclohexane $\Delta H = -119$ kJ mol^{-1}

'cyclohexatriene' + $3H_2$ (g) → cyclohexane
(theoretical compound)

Predicted value
$\Delta H = 3 \times -119$ kJ mol^{-1}
$= -357$ kJ mol^{-1}

benzene + $3H_2$ (g) → cyclohexane $\Delta H = -207$ kJ mol^{-1}

The hydrogenation of benzene releases 150 kJ less energy per mol than the predicted value for cyclohexatriene. This energy is called the **delocalisation energy** or **resonance energy**.

10.2 Functional group chemistry
Syllabus
Nature of science – can you relate this topic to these concepts?
Use of data – much of the progress that has been made to date in the developments and applications of scientific research can be mapped back to key organic chemical reactions involving functional group interconversions.
Understandings – how well can you explain these statements?
Alkanes have low reactivity and undergo free-radical substitution reactions.
Alkenes are more reactive than alkanes and undergo addition reactions. Bromine water can be used to distinguish between alkenes and alkanes.
Alcohols undergo nucleophilic substitution reactions with acids (also called esterification or condensation) and some undergo oxidation reactions.
Halogenoalkanes are more reactive than alkanes. They can undergo (nucleophilic) substitution reactions. A nucleophile is an electron-rich species containing a lone pair that it donates to an electron-deficient carbon.
Addition polymers consist of a wide range of monomers and form the basis of the plastics industry.
Benzene does not readily undergo addition reactions but does undergo electrophilic substitution reactions.
Applications and skills – how well can you do all of the following?
Writing equations for the complete and incomplete combustion of hydrocarbons. Explanation of the reaction of methane and ethane with halogens in terms of a free-radical substitution mechanism involving photochemical homolytic fission.
Writing equations for the reactions of alkenes with hydrogen and halogens and of symmetrical alkenes with hydrogen halides and water. Outline of the addition polymerization of alkenes. Relationship between the structure of the monomer to the polymer and repeating unit.
Writing equations for the complete combustion of alcohols. Writing equations for the oxidation reactions of primary and secondary alcohols (using acidified potassium dichromate(VI) or potassium manganate(VII) as oxidizing agents). Explanation of distillation and reflux in the isolation of the aldehyde and carboxylic acid products. Writing the equation for the condensation reaction of an alcohol with a carboxylic acid, in the presence of a catalyst (eg concentrated sulfuric acid) to form an ester.
Writing the equation for the substitution reactions of halogenoalkanes with aqueous sodium hydroxide.

Alkanes – introduction

Alkanes are **saturated** hydrocarbons with sp^3 hybridization on the carbon single bonds.

Alkanes – physical properties

Alkanes have weak van der Waals forces between the molecules. Boiling point and melting point increases with the M_r of the molecule as the instantaneous induced dipole-induced dipole attraction increases. The first four alkanes are gases at room temperature.

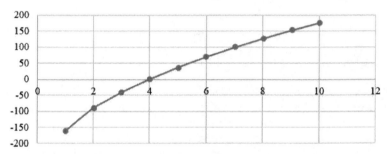

Trend of boiling point in °C for straight chain alkanes with increasing no. of carbon atoms

The effect of branching on boiling point can be illustrated using pentane and its isomers. Boiling point decreases as there are fewer points of contact between adjacent molecules and hence weaker intermolecular forces.

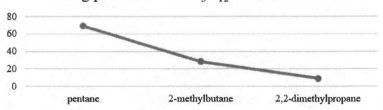

Boiling points in °C of C_5H_{12} structural isomers

Alkanes – Combustion

The major use of alkanes is as fuels because of the large amounts of energy released in combustion.
In a **plentiful supply of air (excess oxygen), carbon dioxide** and **steam** are the products.
Examples:
$CH_4 (g) + 2O_2 (g) \rightarrow CO_2 (g) + 2H_2O (g)$ $\Delta H = -891$ kJ mol^{-1}
$C_3H_8 (g) + 5O_2 (g) \rightarrow 3CO_2 (g) + 4H_2O (g)$ $\Delta H = -2219$ kJ mol^{-1}
$C_8H_{18} (l) + 12½O_2 (g) \rightarrow 8CO_2 (g) + 9H_2O (g)$ $\Delta H = -5470$ kJ mol^{-1}
The reactions are highly exothermic (ΔH negative). Carbon dioxide is a greenhouse gas and contributes to global warming as its polar bonds absorb infrared radiation.
In a **limited supply of air (oxygen), carbon monoxide** and **carbon** are more likely products, in addition to steam. **Hydrogen** is always oxidised (hydrogen is **never a product** of combustion).
$2C_3H_8 (g) + 7O_2 (g) \rightarrow 6CO (g) + 8H_2O (g)$
$C_3H_8 (g) + 2O_2 (g) \rightarrow 3C (s) + 4H_2O (g)$
Carbon monoxide is a poisonous gas. It has a much higher affinity than oxygen with haemoglobin in the blood. This prevents oxygen molecules from being transported by the circulatory system.
Carbon particulates (soot) appears as black smoke and is harmful to the respiratory system. It also causes 'global dimming' – the smog restricts the transmission of sunlight onto the Earth's surface which can be catastrophic to photosynthetic plants.

Alkanes – free-radical substitution (halogenation)

One atom or group is replaced by another atom or group. There are always two reactants and two products. In the presence of bright white or ultraviolet light, one or more of the hydrogen atoms in an alkane can be replaced by Cl or Br (halogenation).

$CH_4 (g) + Cl_2 (g) \rightarrow CH_3Cl (g) + HCl (g)$
$CH_3 (g) + Cl_2 (g) \rightarrow CH_2Cl_2 (g) + HCl (g)$
$C_6H_{14} (l) + Br_2 (l) \rightarrow C_6H_{13}Br (l) + HBr (g)$

Step 1: initiation
In the presence of UV radiation, the halogen molecule undergoes homolytic fission (photolysis) to form two radicals, each one retaining one of the bonding electrons from the covalent bond.

$:\ddot{C}l:\ddot{C}l: \rightarrow :\ddot{C}l\cdot + \cdot\ddot{C}l:$

Step 2: propagation
Free radicals have a short lifespan and are very reactive. Each step pf propagation creates a new free radical that attacks the next species.

$:\ddot{C}l\cdot + H:CH_3 \rightarrow H:\ddot{C}l: + \cdot CH_3$

   chlorine radical attacks   methyl radical and
     methane molecule   hydrogen chloride produced

$\cdot CH_3 + :\ddot{C}l:\ddot{C}l: \rightarrow :\ddot{C}l:CH_3 + :\ddot{C}l\cdot$

   methyl radical attacks   chlorine radical and
     chlorine molecule   chloromethane produced

As the concentration of CH₃Cl increases, an alternative propagation step becomes likely. These reactions use up a free radical as well as produce a free radical so that the reactions can continue as a chain reaction.

$:\ddot{C}l\cdot + :\ddot{C}l:CH_3 \rightarrow H:\ddot{C}l: + \cdot CH_2Cl$

$\cdot CH_2Cl + :\ddot{C}l:\ddot{C}l: \rightarrow :\ddot{C}l:CH_2Cl + :\ddot{C}l\cdot$

Step 3: termination
As the concentration of the hydrocarbon decreases, the radicals attack each other to form a molecule, ending the propagation.

:Cl· + ·Cl: → :Cl:Cl:

:Cl· + ·CH$_3$ → :Cl:CH$_3$

H$_3$C· + ·CH$_3$ → H$_3$C:CH$_3$

Alkenes – introduction

Alkenes are **unsaturated** hydrocarbons with at least one carbon-carbon bond within the molecule. This C=C double bond is the functional group of alkenes known as the alkenyl group, which is more reactive than the C-C single bonds in alkanes.

The main reaction to consider is addition and often involves the simple combination of an alkene with molecules like water or hydrogen. There are four common examples:
- Hydrogen
- Halogens
- Hydrogen halides
- Water (steam)

Alkenes – addition with hydrogen

Alkenes undergo addition reactions with hydrogen gas to form C-C single bonds.
Conditions: temperature of 150°C
Catalyst: finely divided nickel
This reaction is used to convert unsaturated oils to saturated edible fats e.g. palm oil to margarine to give it the following properties:
- solid at room temperature (increased melting point)
- spreadable
- longer shelf life

The partial hydrogenation of polyunsaturated oils have been linked to health risks as the process also converts cis C=C bonds to trans C=C bonds. These trans fats increase the low-density lipoprotein (LDL) concentration and decrease the high-density lipoprotein (HDL) concentration in the blood. LDLs are involved in the transport of cholesterol in the blood which increases the risks of heart disease.

$$C_2H_4\ (g) + H_2\ (g) \xrightarrow[\Delta]{Ni} C_2H_6\ (g)$$

Alkenes – addition with halogens

Alkenes undergo addition reactions with halogens to form dihalogenated alkanes.
Conditions:
For chlorination, chlorine gas is bubbled through the alkene.
For bromination, the alkene is bubbled through liquid bromine or bromine solution.
This decolourisation of bromine is a good test for the presence of a C=C double bond.

$C_2H_4 (g) + Br_2 (l) \rightarrow C_2H_4Br_2 (g)$

$C_4H_8 (g) + Br_2 (l) \rightarrow C_4H_8Br_2 (g)$

Alkenes – addition with halogens (mechanism)

The proximity of the C=C double bond to the bromine molecule induces a dipole moment.
One bromine atom adds to the double bond while the other gains the electron to form a bromide anion.

The bromide anion attacks the carbocation.

→
```
      Br  H
      |   |
  H—C—C—Br
      |   |
      H   H
```

Alkenes – addition with hydrogen halides

Alkenes undergo addition reactions with hydrogen halides to form mono-halogenated alkanes.
Conditions: alkene bubbled through acid solutions of HCl, HBr or HI at room temperature.

C_2H_4 (g) + HBr (aq) → C_2H_5Br (g)

```
  H         H                              H   H
   \       /                               |   |
    C==C       +   H—Br   →    H—C—C—Br
   /       \                               |   |
  H         H                              H   H
```

C_4H_8 (g) + HBr (aq) → C_4H_9Br (g)

```
        H       H                          H   H   Br  H
         \     /                           |   |   |   |
          C==C                             |   |   |   |
         /     \         + H—Br  →  H—C—C—C—C—H
      H /       \ H                        |   |   |   |
       C         C                         H   Br  H   H
      / \       / \
     H   H     H   H
```

Alkenes – addition with water (steam)

Alkenes undergo addition reactions with steam to form alcohols.
Conditions: alkene is mixed with steam at a temperature of 300°C and a pressure of 6-7 MPa.
Catalyst: phosphoric(V) acid, H_3PO_4

C_2H_4 (g) + H_2O (g) → C_2H_5OH (l)

Alcohol is commonly used as a solvent for organic compounds and is now added to petrol to create a biofuel with typically 10% alcohol.

Alkenes – addition polymerisation

When an alkene undergoes addition with itself a process of polymerisation takes place. A long molecular chain builds up.

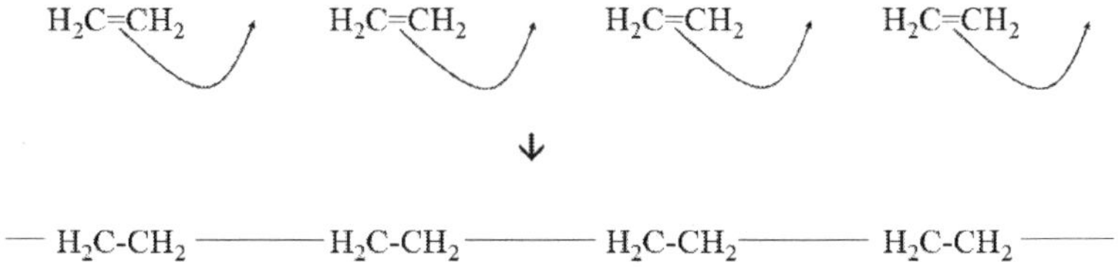

$H_2C=CH_2$, $H_2C=CH_2$, $H_2C=CH_2$, $H_2C=CH_2$,

↓

— H_2C-CH_2 ——— H_2C-CH_2 ——— H_2C-CH_2 ——— H_2C-CH_2 ———

General equation for representing addition polymerisation:

$$n \begin{pmatrix} H & H \\ \diagdown & \diagup \\ C = C \\ \diagup & \diagdown \\ H & X \end{pmatrix} \rightarrow \begin{pmatrix} H & H \\ | & | \\ -C-C- \\ | & | \\ H & X \end{pmatrix}_n$$

monomer → polymer

By substituting X with a suitable group, a whole range of polymers can be made.

X	Repeating unit	Name of polymer
H	$-(CH_2-CH_2)_n-$	polyethene
CH_3	$-[CH_2-CH(CH_3)]_n-$	polypropene
Cl	$-(CH_2-CHCl)_n-$	polychloroethene
C_6H_5	$-[CH_2-CH(C_6H_5)]_n-$	polystyrene
all replaced by F	$-(CF_2-CF_2)_n-$	polytetrafluoroethene

cohols – introduction

Alcohols contain the hydroxyl –OH functional group. The presence of this group enables alcohols to form intermolecular hydrogen bonding with itself and water. As a result:
- alcohols are liquids at room temperature.
- alcohols are miscible with water – the solubility decreases as the hydrocarbon chains increase in length.

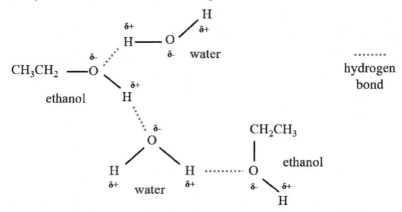

Alcohols – combustion

Alcohols undergo complete combustion in the presence of sufficient oxygen to produce carbon dioxide and water. The lighter alcohols such as methylated spirit are good fuels. It burns with a clean blue flame and is volatile so can be dangerous to use.
Ethanol and methanol are often used as an additive to petrol.
This reduces the carbon footprint of the fuel as the alcohols can be manufactured by fermentation.
The larger the alcohol molecule, the higher the energy output per mole from burning the alcohol due to the increasing alcohol:CO_2 ratio.

Methanol: $2CH_3OH\ (l) + 3O_2\ (g) \rightarrow 2CO_2\ (g) + 4H_2O\ (l)$
$\Delta H = -726$ kJ mol^{-1} (alcohol:CO_2 = 1:1)

Ethanol: $C_2H_5OH\ (l) + 3O_2\ (g) \rightarrow 2CO_2\ (g) + 3H_2O\ (l)$
$\Delta H = -1367$ kJ mol^{-1} (alcohol:CO_2 = 1:2)

Propanol: $2C_3H_7OH\ (l) + 9O_2\ (g) \rightarrow 6CO_2\ (g) + 8H_2O\ (l)$
$\Delta H = -2021$ kJ mol^{-1} (alcohol:CO_2 = 1:3)

Butanol: $C_4H_9OH\ (l) + 6O_2\ (g) \rightarrow 4CO_2\ (g) + 5H_2O\ (l)$
$\Delta H = -2676$ kJ mol^{-1} (alcohol:CO_2 = 1:4)

Pentanol: $2C_5H_{11}OH\ (l) + 15O_2\ (g) \rightarrow 10CO_2\ (g) + 12H_2O\ (l)$
$\Delta H = -3330$ kJ mol^{-1} (alcohol:CO_2 = 1:5)

Alcohols – oxidation of primary alcohols to aldehydes (mild conditions)

Oxidising agent: acidified potassium dichromate(VI), $K_2Cr_2O_7$
- The presence of Cr^{6+} gives it a bright orange colour.
- As $Cr_2O_7^{2-}$ is reduced, the green colour of Cr^{3+} appears.
- The characteristic smell of rotting apples is evident as ethanal (an aldehyde) is formed.

Conditions:
- simple distillation.
- the receiving flask with ethanal should be cooled by an ice bath to prevent evaporation as ethanal has a very low boiling point of 21°C.

$CH_3CH_2OH \rightarrow CH_3CHO$

Alcohols – further oxidation of primary alcohols to carboxylic acids (strong conditions – reflux)

Further oxidation of ethanol is achieved by refluxing it with excess of acidified dichromate(VI) solution for 15 minutes. The reaction mixture which evaporates is recondensed to allow further reaction.
The product is then distilled off and has the characteristic smell of vinegar as ethanoic acid is formed.

$CH_3CHO \rightarrow CH_3COOH$

Alcohols – oxidation of secondary alcohols to ketones (strong conditions – reflux)

Secondary alcohols are oxidised to ketones. The ketones are not oxidised any further as the C=O functional group does not contain any hydrogen atoms.

$CH_3CH(OH)CH_3 \rightarrow CH_3COCH_3$

Alcohols – oxidation of tertiary alcohols

Tertiary alcohols do not react under these conditions as the carbon skeleton requires a lot more energy to break.

Alcohols – oxidation (summary)

primary alcohols $\xrightarrow{\text{mild distillation}}$ aldehydes $\xrightarrow{\text{strong reflux}}$ carboxylic acids

secondary alcohols $\xrightarrow{\text{strong reflux}}$ ketones

tertiary alcohols $\longrightarrow$ no change

Examples:

$CH_3CH_2CH_2OH$ (l) $\longrightarrow$ CH_3CH_2CHO (l) $\longrightarrow$ CH_3CH_2COOH (aq)
  propan-1-ol   propanal   propanoic acid

$(CH_3)_2CHOH$ (l) $\longrightarrow$ CH_3COCH_3 (l)
  propan-2-ol   propanone

Alcohols – condensation with carboxylic acid to form esters

When two compounds with functional groups that react with each other react, this is known as a condensation reaction. A simple molecule e.g. H_2O, NH_3 or HCl is eliminated with each link in a condensation reaction.
Examples:
A-OH + H-B → A-B + H_2O
A-H + Cl-B → A-B + HCl
A-H + H_2N-B → A-B + NH_3

Carboxylic acids react with **alcohols** when **warmed under reflux** together in the presence of **concentrated sulfuric acid (catalyst)**. The reaction is **reversible** and the product is an **ester** – the -COO- functional group is called the ester group.

carboxylic acid + alcohol $\rightleftharpoons$ ester + water
If the reaction mixture is **poured into cold water** and the excess **carboxylic acid neutralised with sodium hydrogencarbonate**, the characteristic sweet smell is immediately recognisable.

RCOO is the **carboxylate group** from the original **carboxylic acid**.
R' is the **alkyl or aryl** group from the original **alcohol**.

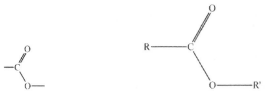

Esters have **sweet smells** and many are naturally occurring substances and are used as **artificial flavourings**. Examples:

$CH_3CH_2CH(CH_3)COOCH_2CH_3$
ethyl 2-methylbutanoate
apple flavour

$CH_3COOCH_2CH_2CH(CH_3)CH_3$
3-methylbutyl ethanoate
ripe pears

$CH_3COOCH(CH_3)CH_2CH_2CH_3$
1-methylbutyl ethanoate
banana flavour

$CH_3CH_2CH_2COOCH_2CH_2CH_2CH_3$
butyl butanoate
pineapple flavour

When naming esters, the alkyl group from the alcohol precedes the carboxylate group.
Examples:

$CH_3COOH\ (l) + C_2H_5OH\ (l) \rightleftharpoons CH_3COOC_2H_5\ (l) + H_2O\ (l)$
ethanoic acid + ethanol $\rightleftharpoons$ ethyl ethanoate + water

$C_3H_7COOH\ (l) + CH_3OH\ (l) \rightleftharpoons C_3H_7COOCH_3\ (l) + H_2O\ (l)$
butanoic acid + methanol $\rightleftharpoons$ methyl butanoate + water

Introduction to mechanisms

Homolytic fission – formation of radicals

Free radicals are atoms or groups of atoms with an unpaired electron that it uses to form a covalent bond. They are formed when a bond breaks by homolytic fission – one electron going to each atom in the bond.

$$:\!\ddot{A}\!:\!\ddot{B}\!: \rightarrow :\!\ddot{A}\!\cdot\; +\; \cdot\ddot{B}\!:$$

Heterolytic fission – formation of ions
This occurs when a bond breaks and both electrons are taken by one of the atoms or groups. The products are ions of opposite charges.

$$:\!\ddot{A}\!:\!\ddot{B}\!: \rightarrow :\!\ddot{A}^{+}\; +\; :\!\ddot{B}\!:^{-}$$

Electrophiles

An **electrophile** is a species that **accepts electron pairs to form a covalent bond** in a reaction or **attacks sites of high electron density**. Electrophiles can be:
- positive ions e.g. H^+, Br^+ or NO_2^+. These could be the positive ion formed during heterolytic fission.
- δ+ atoms in a molecule e.g. the δ+ hydrogen atom in HBr.
- neutral molecules that can be polarised e.g. Br-Br.

Nucleophiles

A **nucleophile** is a species with a lone pair of electrons that it uses to form a covalent bond.
Nucleophiles can be:
- negative ions e.g. OH⁻, Cl⁻, Br⁻, I⁻, CH_3O^- or CN⁻. These could be the negative ion formed during heterolytic fission.
- neutral molecules that contain oxygen or nitrogen atoms e.g. H_2O and NH_3.

Reactions are classified according to the nature of the first attacking agent.
If the **attacking species** is an **electrophile**, the reaction is **electrophilic**. For instance, if the attacking species is the electrophile $Br^{\delta+}$ attacking the electron-rich double of ethene, the mechanism is electrophilic addition.
If the **attacking species** is an **nucleophile**, the reaction is **nucleophilic**. For instance, if the attacking species is the nucleophile :OH⁻ attacking the carbocation, the mechanism is nucleophilic substitution.

Introduction to mechanisms – nucleophilic substitution

In halogenoalkanes, the C-X bond is **polar** but does not have an area of high electron density.
The δ+ carbon atom is attacked by **nucleophiles** such as OH⁻.
C_2H_5Cl (g) + OH⁻ (aq) → C_2H_5OH (aq) + Cl⁻ (aq)

Introduction to mechanisms – electrophilic substitution

In benzene, the ring is an area of high electron density and is susceptible to attacks by electrophiles such as cations.
Benzene reacts with bromine, Br_2 in the presence of a Lewis base catalyst, $FeBr_3$ reacts to form bromobenzene.

⌬ + Br_2 → ⌬-Br + HBr

20.1 Types of organic reactions
Syllabus
Nature of science – can you relate this topic to these concepts?
Looking for trends and discrepancies—by understanding different types of organic reactions and their mechanisms, it is possible to synthesize new compounds with novel properties which can then be used in several applications. Organic reaction types fall into a number of different categories. Collaboration and ethical implications—scientists have collaborated to work on investigating the synthesis of new pathways and have considered the ethical and environmental implications of adopting green chemistry.

Applications and skills – how well can you do all of the following?
Nucleophilic Substitution Reactions:
- Explanation of why hydroxide is a better nucleophile than water.
- Deduction of the mechanism of the nucleophilic substitution reactions of halogenoalkanes with aqueous sodium hydroxide in terms of S_N1 and S_N2 mechanisms. Explanation of how the rate depends on the identity of the halogen (ie the leaving group), whether the halogenoalkane is primary, secondary or tertiary and the choice of solvent.
- Outline of the difference between protic and aprotic solvents.

Electrophilic Addition Reactions:
- Deduction of the mechanism of the electrophilic addition reactions of alkenes with halogens/interhalogens and hydrogen halides.
- Electrophilic Substitution Reactions:
- Deduction of the mechanism of the nitration (electrophilic substitution) reaction of benzene (using a mixture of concentrated nitric acid and sulphuric acid).

Reduction Reactions:
- Writing reduction reactions of carbonyl containing compounds: aldehydes and ketones to primary and secondary alcohols and carboxylic acids to aldehydes, using suitable reducing agents.
- Conversion of nitrobenzene to phenylamine via a two-stage reaction.

Understandings – how well can you explain these statements?
Nucleophilic Substitution Reactions:
- S_N1 represents a nucleophilic unimolecular substitution reaction and S_N2 represents a nucleophilic bimolecular substitution reaction. S_N1 involves a carbocation intermediate. S_N2 involves a concerted reaction with a transition state.
- For tertiary halogenoalkanes the predominant mechanism is S_N1 and for primary halogenoalkanes it is S_N2. Both mechanisms occur for secondary halogenoalkanes.

- The rate determining step (slow step) in an S_N1 reaction depends only on the concentration of the halogenoalkane, rate = k[halogenoalkane]. For S_N2, rate = k[halogenoalkane][nucleophile]. S_N2 is stereospecific with an inversion of configuration at the carbon.
- S_N2 reactions are best conducted using aprotic, non-polar solvents and S_N1 reactions are best conducted using protic, polar solvents.

Electrophilic Addition Reactions:
- An electrophile is an electron-deficient species that can accept electron pairs from a nucleophile. Electrophiles are Lewis acids.
- Markovnikov's rule can be applied to predict the major product in electrophilic addition reactions of unsymmetrical alkenes with hydrogen
- halides and interhalogens. The formation of the major product can be explained in terms of the relative stability of possible carbocations in the reaction mechanism.

Electrophilic Substitution Reactions:
- Benzene is the simplest aromatic hydrocarbon compound (or arene) and has a delocalized structure of π-bonds around its ring. Each carbon to carbon bond has a bond order of 1.5. Benzene is susceptible to attack by electrophiles.

Reduction Reactions:
- Carboxylic acids can be reduced to primary alcohols (via the aldehyde). Ketones can be reduced to secondary alcohols. Typical reducing agents are lithium aluminium hydride (used to reduce carboxylic acids) and sodium borohydride.

Nucleophilic substitution

Quick recap: A **nucleophile** is a species with a lone pair of electrons that it uses to form a covalent bond. Nucleophiles can be:
negative ions e.g. OH^-, Cl^-, Br^-, I^-, CH_3O^- or CN^-. These could be the negative ion formed during heterolytic fission.
neutral molecules that contain oxygen or nitrogen atoms e.g. H_2O and NH_3.
If the **attacking species** is an **nucleophile**, the reaction is **nucleophilic**. For instance, if the attacking species is the nucleophile :OH^- attacking the carbocation, the mechanism is nucleophilic substitution.
In halogenoalkanes, the C-X bond is **polar** but does not have an area of high electron density.
The δ+ carbon atom is attacked by **nucleophiles** such as OH^-.
C_2H_5Cl (g) + OH^- (aq) → C_2H_5OH (aq) + Cl^- (aq)

Halogenoalkanes are very susceptible to nucleophilic substitution reactions. The **highly electronegative halogen atom** causes the adjacent carbon atom to be **deficient of electrons**. This means that it can be easily attacked by a nucleophile, which has a lone pair of electrons to donate.
Nucleophilic substitution can happen via two mechanisms, S_N1 or S_N2, depending on whether the halogenoalkane is primary, secondary or tertiary.

The hydroxide ion, OH⁻ as a nucleophile

Both water, H_2O and the hydroxide ion, OH^- have at least one pair of electrons that it can use to form a covalent bond with an **electrophile**. OH^- is a better nucleophile than H_2O because it has a negative charge and is more strongly attracted to the electrophile.

Rules for drawing chemical mechanisms

- curly arrows start from the lone pair electrons or the covalent bond (bonding pair electrons).
- curly arrows point towards the species being attacked or the leaving group.

Nucleophilic substitution – S_N2 (a one-step reaction for primary halogenoalkanes)

S_N2, which stands for 'nucleophilic substitution reaction of a second order'. Therefore, the molecularity of the reaction is bimolecular as there are two species involved in the initial step of the reaction. The rate of reaction depends on the concentrations of both reactants.
The general rate equation is: **rate = k[halogenoalkane][nucleophile]**
An example would be the reaction between 1-bromobutane and sodium hydroxide to form butan-1-ol.

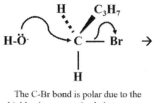 → 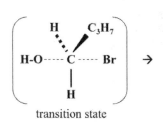 →

The C-Br bond is polar due to the highly electronegative halogen atom.

$\overset{\delta+}{C} \text{—} \overset{\delta-}{Br}$

transition state

OH⁻ attacks the electron deficient carbon atom from the opposite side of the large bromine atom due to the **steric hindrance** of the bromine atom.

The C-OH bond is partially formed and the C-Br bond is partially broken in the transition state.

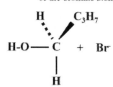 + Br⁻

The C-OH bond is formed on the opposite side of the original C-Br bond and the configuration is inverted. This is known as the **Walden inversion**. Therefore, S_N2 reactions are **stereospecific**.

Nucleophilic substitution – S_N1 (a two-step reaction for tertiary halogenoalkanes)

S_N1, which stands for 'nucleophilic substitution reaction of a first order'. Therefore, the molecularity of the reaction is unimolecular as there is only one species involved in the first step (the slow/rate determining step). The rate of reaction depends on the concentration of the halogenoalkane only.
The general rate equation is: **rate = k[halogenoalkane]**
An example would be the reaction between 2-bromo-2-methyl propane and sodium hydroxide to form 2-methylpropane-2-ol.

$$H_3C-\underset{\underset{CH_3}{|}}{\overset{\overset{CH_3}{|}}{C}}-Br \rightarrow H_3C-\underset{\underset{CH_3}{|}}{\overset{\overset{CH_3}{|}}{C^+}} \quad \ddot{\text{O}}\text{-H} \rightarrow$$

The methyl (-CH_3) groups have a **positive inductive effect** – they push electron density away from themselves. The highly electronegative halogen atom polarises the C-Br bond so the halogen readily leaves the molecule.

carbocation intermediate
The tertiary carbocation has extra stability (compared to 1° or 2° carbocations) as the increased electron density on the central carbon reduces its positive charge.

$$H_3C-\underset{\underset{CH_3}{|}}{\overset{\overset{CH_3}{|}}{C}}-OH \; + \; Br^-$$

$$R \xrightarrow{\quad} C^+ \xleftarrow{\quad} R''$$
$$\underset{R'}{\overset{\uparrow}{|}}$$
movement of electron density due to inductive effect

Nucleophilic substitution – factors determining the rate of reaction

Factor 1: The halogen
The rate determining step in both S_N1 and S_N2 involves the heterolytic fission of the C-X bond. Therefore, the rate of reaction increases with the effectiveness of the halogen as a leaving group.
Although fluorine is the most electronegative and forms the most polar C-X bond, the small radius of fluorine causes the C-F bond to be very short and strong.
The rate of reaction follows the trend: R-I > R-Br > R-Cl > R-F

Factor 2: The class of the halogenoalkane
Tertiary carbocations have extra stability due to the positive inductive effect of the alkyl groups.
- Primary halogenoalkanes mostly undergo nucleophilic substitutions via the S_N2 mechanism.
- Tertiary halogenoalkanes mostly undergo nucleophilic substitutions via the S_N1 mechanism.
- Secondary halogenoalkanes undergo nucleophilic substitutions via both mechanisms.

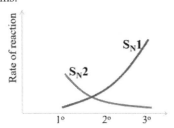

Factor 3: Solvent
Polar, protic solvents e.g. water and alcohols, contain O-H or N-H bonds which form hydrogen bonding with the nucleophile. This makes the nucleophile less nucleophilic as **solvation** occurs – the solvent molecules surround the nucleophile. In the case of water as the solvent, this is called a hydration sphere.
As the nucleophile is now less effective, S_N1 reactions are more favoured.

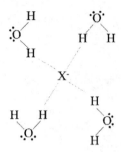

Polar, aprotic solvents e.g. propanone (CH_3COCH_3), ethyl ethanoate ($CH_3COOC_2H_5$) and acetonitrile (CH_3CN), do not contain O-H or N-H bonds and cannot form hydrogen bonding with the nucleophile. This makes the nucleophile more nucleophilic and S_N2 reactions are more favoured.

Electrophilic addition

Quick recap: An **electrophile** is a species that **accepts electron pairs to form a covalent bond** in a reaction or **attacks sites of high electron density**. Therefore, electrophiles act as Lewis acids. Electrophiles can be:
- positive ions e.g. H^+, Br^+, CH_3^+ or NO_2^+. These could be the positive ion formed during heterolytic fission.
- δ+ atoms in a polar molecule e.g. the δ+ hydrogen atom in HBr.
- neutral molecules that can be polarised e.g. Br-Br.

Alkenes are very susceptible to electrophilic addition reactions. The double covalent bond is stronger than a single bond and consists of a sigma (σ) bond and a pi (π) bond. However, it can easily be attacked by electrophiles because:
- it is a site of high electron density.
- the planar shape of the molecule due to sp^2 hybridization, with bond angles of approximately 120° provides less steric hindrance to attacking electrophiles.

Electrophilic addition – alkene and halogen reactions

The halogen molecule is polarised as it approaches the C=C bond. The atom furthest away from the double bond acquires a partial negative charge as the electron density within the halogen molecule is repelled by the electrons in the double bond.

The first halogen atom adds to the carbon atom furthest away from the alkyl group as there is less steric hindrance.

The remaining halide ion acts as a nucleophile and attacks the positively charged carbon atom in the carbocation.

N.B.: Although the electrophile 'attacks' the electron-rich site of the C=C double bond, the curly arrow points from the double bond to the attacking electrophile.

Electrophilic addition – alkene and hydrogen halide reactions

[Mechanism diagram showing propene + HBr → carbocation intermediate → 2-bromopropane]

The polar HX molecule splits heterolytically to form the hydrogen cation, H^+ and the halide anion, X^-, while the H^+ ion attacks the double bond simultaneously.

The charge is then redistributed to form the unstable carbocation. The remaining halide ion acts as a nucleophile and attacks the positively charged carbon atom in the carbocation.

Markovnikov's Rule

When a hydrogen halide adds to an asymmetrical alkene, there are two possible products.

The major product will be the one which follows the route that forms the most stable carbocation. Due to the positive inductive effect of alkyl groups, tertiary carbocations are the most stable as the positive charge on the central carbon atom is reduced.

[Mechanism diagram showing propene + HBr with two pathways: one forming secondary carbocation leading to 2-bromopropane (major), the other forming primary carbocation leading to 1-bromopropane (minor)]

2-bromopropane is the major product as the mechanism involves the formation of a more stable secondary carbocation.

1-bromopropane is the minor product as the mechanism involves the formation of a less stable primary carbocation.

Electrophilic addition – alkene and interhalogen reactions

Interhalogens are diatomic molecules formed from two different halogens. The products formed will depend on the relative electronegativities of the halogens. Markovnikov's rule does not apply in these reactions.

1-bromo-2-chloropropane

The more electronegative halogen is pushed away from the C=C double bond. Therefore, the alkene will be attacked first by the less electronegative of the two halogen atoms.

The remaining halide ion acts as a nucleophile and attacks the positively charged carbon atom in the carbocation.

Electrophilic substitution – nitration of benzene

Benzene undergoes electrophilic substitution reactions instead of addition. The reaction occurs in **two steps**:

1. **Addition**

An electrophile (a cation) attacks the high-electron density delocalised π-ring of benzene. The full delocalisation is broken and a positively charged intermediate is formed. The delocalised ring is very stable and a very high activation energy is needed to break it, so this is the slow (rate determining) step.

2. **Elimination**

A proton (H^+) is lost from the ring and an electron is returned to the ring to complete the delocalisation. The aromaticity is restored for extra stability as opposed to an addition of an anion (as in the reaction with alkenes).

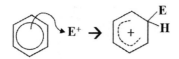

The overall reaction for the formation of nitrobenzene can be shown as:

$$C_6H_6 + HNO_3 \xrightarrow[50°C]{H_2SO_4} C_6H_5NO_2 + H_2O$$

The product is a yellow oil and can be separated from the aqueous reaction mixture.

Electrophilic substitution – nitration of benzene

Generation of the electrophile, NO_2^+ (nitronium cation)

A high concentration of nitronium cations can be produced from a nitrating mixture of sulfuric and nitric acids at 50°C.

[Structure of nitric acid] + H_2SO_4 ⇌ [protonated nitric acid] + HSO_4^-

Nitric acid is protonated by sulfuric acid.

Release of H_2O forms the strong electrophile, NO_2^+.

⇌ $O=N^+=O$ + H_2O + HSO_4^-
   nitronium ion

Addition

[benzene] + NO_2^+ → [cyclohexadienyl cation with NO_2 and H]

Rate determining step: Benzene ring donates a pair of protons to form a new C-N bond. π-electron from N=O bond moves onto oxygen atom.

Elimination

[carbocation intermediate with NO_2 and H] → [nitrobenzene with NO_2] + H^+

H_2O deprotonates the carbocation intermediate and restores the delocalization in the ring.

Synthesis of phenylamine (aniline) from nitrobenzene

Nitrobenzene can be converted to phenylamine (aniline) in a two-step reaction using concentrated hydrochloric acid and zinc, followed by sodium hydroxide:

Step 1: Nitrobenzene is heated under reflux in a water bath for about half an hour with a mixture of concentrated hydrochloric acid and zinc.

$$C_6H_5NO_2 \text{ (l)} + 3Zn \text{ (s)} + 7H^+ \text{ (aq)}$$

$$\rightarrow C_6H_5NH_3^+ \text{ (aq)} + 3Zn^{2+} \text{ (s)} + 2H_2O \text{ (l)}$$

Phenylammonium ion is formed. Zn is oxidized to Zn^{2+}.

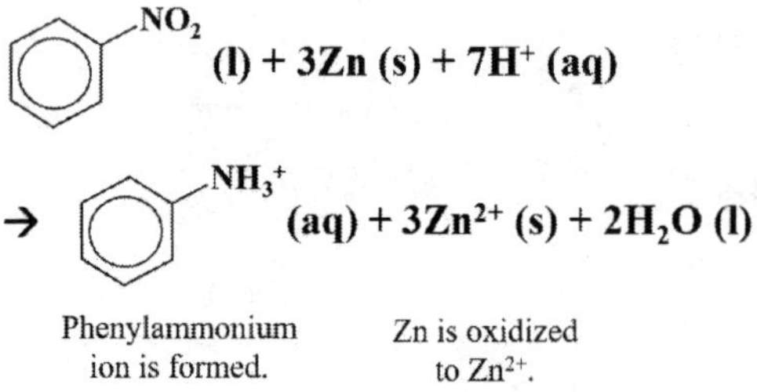

Step 2: Sodium hydroxide (NaOH) is added

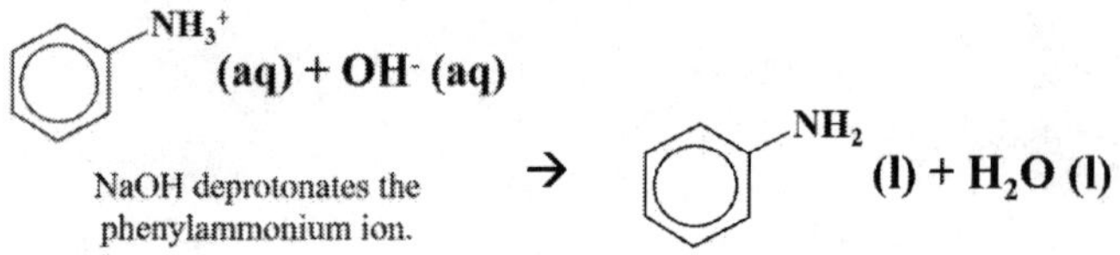

$$C_6H_5NH_3^+ \text{ (aq)} + OH^- \text{ (aq)}$$

NaOH deprotonates the phenylammonium ion.

$$\rightarrow C_6H_5NH_2 \text{ (l)} + H_2O \text{ (l)}$$

Oxidation of alcohols and aldehydes (review)

primary alcohols $\xrightarrow{\text{mild}}_{\text{distillation}}$ aldehydes $\xrightarrow{\text{strong}}_{\text{reflux}}$ carboxylic acids

secondary alcohols $\xrightarrow{\text{strong}}_{\text{reflux}}$ ketones

tertiary alcohols $\longrightarrow$ no change

Nucleophilic addition – reduction of carboxylic acids, ketones and aldehydes

The above reactions can be 'reversed' by using a suitable reducing agent that can react with the polar π-bond in the C=O groups:
- sodium borohydride, $NaBH_4$ can reduce aldehydes to primary alcohols and ketones to secondary alcohols. [H] denotes the reducing agent.

R-CHO + 2[H] → RCH_2-OH
R-CO-R' + 2[H] → R-CH(OH)-R'

- lithium aluminium hydride, $LiAlH_4$ is a stronger reducing agent and can reduce carboxylic acids to primary alcohols.

20.2 Synthetic routes
Syllabus
Nature of science – can you relate this topic to these concepts?
Scientific method—in synthetic design, the thinking process of the organic chemist is one which invokes retro-synthesis and the ability to think in a reverse-like manner.
Applications and skills – how well can you do all of the following?
Deduction of multi-step synthetic routes given starting reagents and the product(s).
Understandings – how well can you explain these statements?
The synthesis of an organic compound stems from a readily available starting material via a series of discrete steps. Functional group interconversions are the basis of such synthetic routes.
Retro-synthesis of organic compounds.

Synthesis of organic compounds

The production of organic compounds is an important process in the production of many substances including medicines, fertilisers, food additives and many others. Therefore, the finished product is most profitable when:
the process involves the fewest steps.
the yield in each step is high.
For reactions involving an equilibrium, the conditions are optimised to produce the best yield in the shortest possible time. The temperature, pressure and the use of catalysts must be carefully controlled.
Retro-synthesis
Many substances are found in nature e.g. plant-based medicines but these are often very expensive to extract from biological material. Scientists often produce these chemicals in the laboratory by retro-synthesis. This is the process of breaking down a large, complex molecules into smaller constituents which can be produced in a laboratory under conditions which are economically viable. These steps can involve the growth or shortening of molecular chains or the interconversion of organic functional groups.

Reaction map

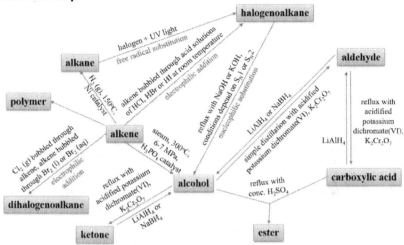

20.3 Stereoisomerism
Syllabus
Nature of science – can you relate this topic to these concepts?
Transdisciplinary—the three-dimensional shape of an organic molecule is the foundation pillar of its structure and often its properties. Much of the human body is chiral.
Applications and skills – how well can you do all of the following?
- Construction of 3-D models (real or virtual) of a wide range of stereoisomers.
- Explanation of stereoisomerism in non-cyclic alkenes and C_3 and C_4 cycloalkanes.
- Comparison between the physical and chemical properties of enantiomers.
- Description and explanation of optical isomers in simple organic molecules.
- Distinction between optical isomers using a polarimeter.

Understandings – how well can you explain these statements?
- Stereoisomers are subdivided into two classes—conformational isomers, which interconvert by rotation about a σ bond and configurational isomers that interconvert only by breaking and reforming a bond.
    - Isomerism is divided into - Stereoisomerism & Structural Isomerism
    - Stereoisomerism is further divided into - Configurational Isomerism & Conformational Isomerism
    - Configurational isomerism is further divided into - cis-trans and E/Z isomerism and optical isomerism.
- Cis-trans isomers can occur in alkenes or cycloalkanes (or heteroanalogues) and differ in the positions of atoms (or groups) relative to a reference plane.
- According to IUPAC, E/Z isomers refer to alkenes of the form $R_1R_2C=CR_3R_4$ ($R_1 \neq R_2$, $R_3 \neq R_4$) where neither R_1 nor R_2 need be different from R_3 or R_4.
- A chiral carbon is a carbon joined to four different atoms or groups.
- An optically active compound can rotate the plane of polarized light as it passes through a solution of the compound. Optical isomers are enantiomers. Enantiomers are non-superimposeable mirror images of each other. Diastereomers are not mirror images of each other.
- A racemic mixture (or racemate) is a mixture of two enantiomers in equal amounts and is optically inactive.

Isomerism – a review

Isomers are compounds with the same molecular formula but different structural formulae (different arrangement of atoms).

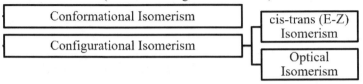

Corformational Isomerism

Corformational isomerism occurs when the groups attached to a C-C single interconvert by rotation about the σ-bond without breaking it. The structures of conformational isomers can be shown with Newman projections.

3-D representation

Newman projections

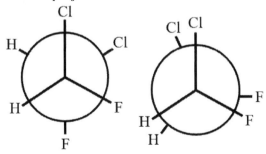

Staggered conformation: the groups attached to the adjacent carbon atoms are positioned at 60° to each other – they are as far away from each other as possible.
- more stable.
- less repulsion between the electrons in the bonds opposite to each other.

Eclipsed conformation: the groups attached to the adjacent carbon atoms are positioned at 0° to each other. This is shown askew for clarity.
- less stable (high energy).
- more repulsion between the electrons in the bonds opposite to each other (high torsional strain).

Configurational Isomerism – cis-trans and E-Z isomerism

Configurational isomerism occurs when bonds are broken (cis-trans or E-Z) or the stereocentres of chiral molecules are rearranged (optical isomerism). Cis-trans or E-Z isomerism occur when there is restricted rotation about the bonds – in planar or ring (cyclic) molecules.
cis-isomers: substituents are on the same side of the reference plane.
trans-isomers: substituents are on opposite sides of the reference plane.

Planar molecules
- the π-bond in the C=C double bond restricts rotation.
- bond angles around each carbon atom in the double bond is about 120°.

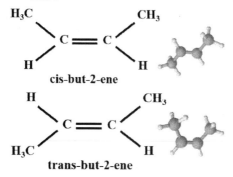

The chemotherapy drug, cisplatin is used as a treatment for cancer. The complex molecule inhibits cell division in cancer cells. It also causes hair loss as it stops the regrowth of new hair.
The trans-isomer has no biological activity.

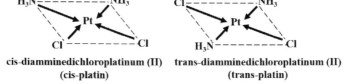

Cyclic molecules
- all the carbon atoms in the ring are bonded together and restricts rotation.

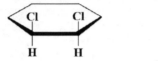
cis-1,2-dichlorocyclohexane

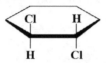
trans-1,2-dichlorocyclohexane

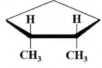
cis-1,2-dimethylcyclopentane

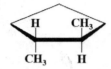
trans-1,2-dimethylcyclopentane

cis-1,2-dimethylcyclobutane

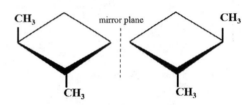
trans-1,2-dimethylcyclobutane

The cis-trans system of naming geometric isomers does not work for compounds with three or more substituents. The substituents are ordered in priority by the atomic number (Z) of the atom directly bonded to the carbon atom according to the Cahn-Ingold-Prelog (CIP priority) system.

Higher priority substituents on opposite sides (Entgegen)
Higher priority substituents on together (Zusammen)

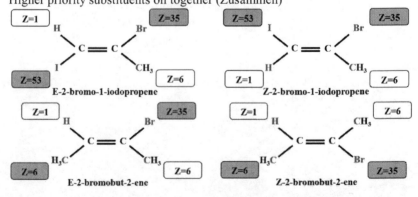

Configurational Isomerism – optical isomerism

The simplest form of optical isomerism is where a carbon atom is joined to four different groups. The groups can be arranged in two different ways. The two isomers are mirror images of each other which cannot be superimposed e.g. the left and right hands are an example of non-superimposable mirror images.

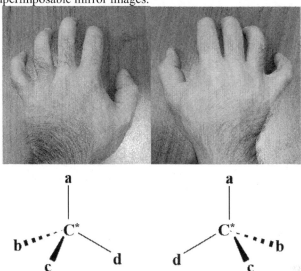

The two isomers are known as **chiral** (Greek for 'hand')molecules. The carbon atom carrying the 4 different groups is a **chiral centre (stereocentre** or **asymmetric centre)**, shown as C*. Optical isomerism is often called '**chirality**'.

Examples of optical isomerism include 2-amino acids:

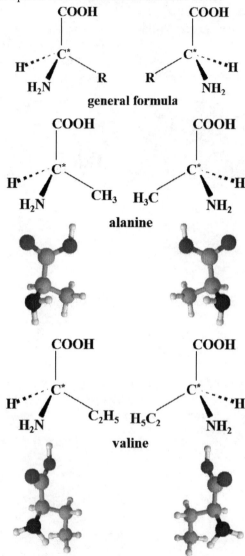

general formula

alanine

valine

Note that the smallest amino acid, glycine is not optically active as there are two hydrogen atoms bonded to the central carbon i.e. it is not bonded to four different groups.

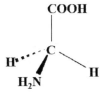

Molecules that are optical isomers to each other are called **enantiomers** or **diastereomers**. The enantiomers will have identical physical properties e.g. melting/boiling points, solubility, density and viscosity but can be distinguished from each other because:
- they rotate the plane of plane-polarised light in opposite directions.
- they react differently with other chiral molecules.

The angle of rotation of the plane-polarised light depends on:
- the nature of the enantiomer.
- the concentration of the enantiomer in the solution.

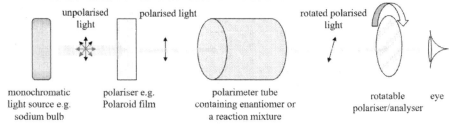

Some chemical reactions result in a 50:50 mixture of the two enantiomers. These are called **racemic mixtures** and have no effect on plane-polarised light e.g. tartaric acid.

(+)-tartaric acid — clockwise rotation of polarised light

(-)-tartaric acid — anticlockwise rotation of polarised light

Chiral molecules often have distinct physical properties and react differently with other chiral molecules – the molecule must be the correct shape to fit the molecule it is reacting with.

Many natural molecules e.g. amino acids (and so proteins) are chiral and most natural reactions are affected by just one of the enantiomers. All naturally occurring amino acids are levorotary (-). Many drugs are also optically active with one enantiomer being beneficial and the other being inert or even harmful. When synthesised in the laboratory, chiral molecules often exist as a racemic mixture.

In the 1950s thalidomide was given to pregnant women to reduce the effects of morning sickness. This led to many birth defects and early deaths in many cases. Thalidomide was banned worldwide when the effects were discovered. However, it is starting to be used again to treat leprosy and HIV.

Its use is restricted though and patients have to have a pregnancy test first (women!) and use two forms of contraception (if sexually active). The body racemises each enantiomer, so even the effective enantiomer is dangerous as it converts to the harmful one in the body.

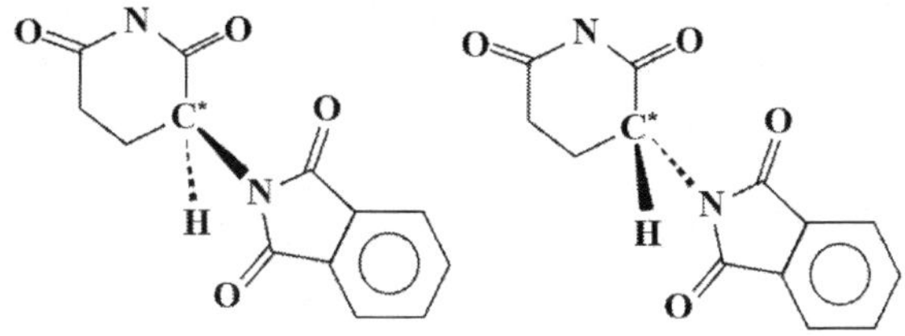

the two enantiomers of thalidomide

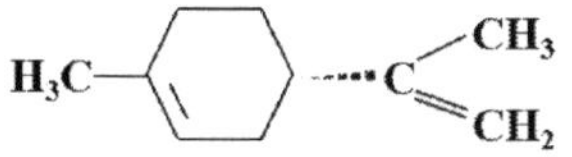

star anise and pine needles

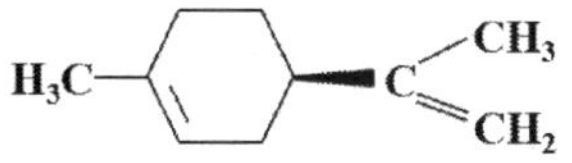

oranges and lemons

the two enantiomers of limonene

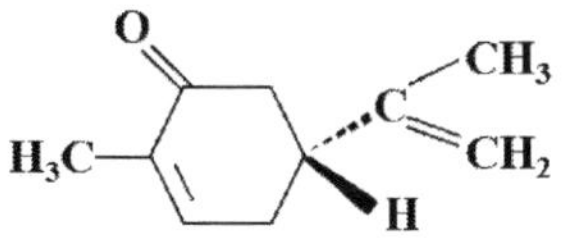

caraway seeds

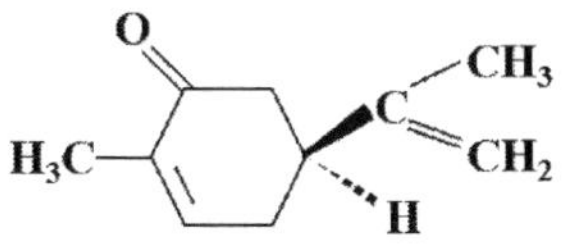

spearmint

the two enantiomers of carvone

Diastereomers are a type of a stereoisomer, when two or more stereoisomers of a compound have different configurations at one or more (but not all) of the equivalent (related) stereocentres and are not mirror images of each other.

Due to the more complex interactions between the different groups, diastereomers differ in both physical and chemical properties.

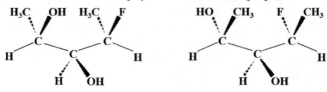

11.1 Uncertainties and errors in measurement and results
Syllabus
Nature of science – can you relate this topic to these concepts?
Making quantitative measurements with replicates to ensure reliability – precision, accuracy, systematic, and random errors must be interpreted through replication.

Understandings – how well can you explain these statements?
Qualitative data includes all non-numerical information obtained from observations not from measurement.

Quantitative data are obtained from measurements, and are always associated with random errors/uncertainties, determined by the apparatus, and by human limitations such as reaction times.

Propagation of random errors in data processing shows the impact of the uncertainties on the final result.

Experimental design and procedure usually lead to systematic errors in measurement, which cause a deviation in a particular direction.

Repeat trials and measurements will reduce random errors but not systematic errors.

Applications and skills – how well can you do all of the following?
Distinction between random errors and systematic errors.

Record uncertainties in all measurements as a range (±) to an appropriate precision.

Discussion of ways to reduce uncertainties in an experiment.

Propagation of uncertainties in processed data, including the use of percentage uncertainties.

Discussion of systematic errors in all experimental work, their impact on the results and how they can be reduced.

Estimation of whether a particular source of error is likely to have a major or minor effect on the final result.

Calculation of percentage error when the experimental result can be compared with a theoretical or accepted result.

Distinction between accuracy and precision in evaluating results

Significant figures (SF)

Numbers greater than 1: Include ALL the numbers except trailing zeros which are not follow by a decimal point.

Number	102	1.02	1.020	10.20	100
SF	3	3	4	4	1

Numbers smaller than 1: Include ALL the numbers except the zeros before and immediately after the decimal point.

Number	0.202	0.2020	0.00202	0.02200
SF	3	4	3	4

Scientific notation: Ignore the power of 10.

Number	1.02×10^{-7}	7.623×10^{12}	4.60×10^{-5}
SF	3	4	3

Multiplication and **division**: the result should have the same number of **significant figures** as the least precise piece of data.
Addition and **subtraction**: the result should not have more **decimal places** than the least precise piece of data.
To round off numbers to significant figures, look for the next significant figure i.e.
< 5 : keep the same number.
≥ 5 : round up the number.

Example 1:

26.09 g of potassium chloride (KCl) was needed to prepare 100.00 cm^3 of saturated solution. What is the concentration of this solution?
amount of KCl = 26.09 / (39.10 + 35.45) = 0.349966 mol ≈ 0.3500 mol
volume of H$_2$O = 100.00 cm^3 = 100.00 / 1000 = 0.10000 dm^3
concentration of KCl solution = 0.3500 / 0.10000 = 3.500 mol dm^{-3}

Example 2:
A sample of uranium fluoride with a mass of 8.637 g contains 5.84 g uranium.
A_r (U) = 238.03 g mol^{-1} A_r (F) = 19.00 g mol^{-1}

Data	Calculation	Incorrect	Correct	Reason
Mass of fluorine	8.637-5.84	2.797 g	2.80 g	Only 2 d.p. in the mass data for uranium
Moles of uranium	5.84/238.03	0.0245347 mol	0.0245 mol	Only 3 s.f. in the mass data for uranium
Moles of fluorine	2.797/19.00	0.1472105 mol	0.147 mol	Mass of fluorine is from the first calculation
F:U ratio	0.147/0.0245	6.0000937	6.00	

Experimental errors – systematic errors

- affect a result in a **particular direction** only.
- arise from flaws or defects in the instrument or from errors in the way that the measurement was taken.
- difficult to allow for quantitatively but the direction in which it would affect the final result can always be determined.

Examples of systematic errors:
- taking the mass of an empty weighing bottle rather than re-weighing the bottle after tipping out its contents.
- not reading the scale of a burette at eye level (parallax error).
- reading the top of the meniscus in measuring cylinders or pipettes.
- wrongly calibrating instruments such as pH meters.
- heat exchange (loss) to the surroundings in calorimetric experiments.

Experimental errors – random errors

- arise from **uncontrolled variation** in measurements of variables.
- make a measurement **less precise**, but not in any particular direction – the actual value may be either greater or smaller than the value that is recorded.
- cannot be eliminated but **can be reduced by repeat measurements** – the random variations cancel out statistically.

Examples of random errors:
- reaction time.
- sudden changes in room temperature when taking thermometer readings.
- accumulation of dust on top pan balance while measuring mass.

Accuracy and Precision

Accuracy (of measurement): Closeness of agreement between the result of a measurement and a true value of the particular quantity being measured. A result which is accurate is one which is **close to the true (accepted) value**.
Precision: The closeness of agreement between independent test results obtained by applying the experimental procedure under stipulated conditions. The smaller the random part of the experimental errors which affect the results, the more precise the procedure. A measure of precision (or imprecision) is the standard deviation. A precise instrument gives a **consistent reading** when it is used repeatedly for the same measurements. The uncertainty of the reading or result will be smaller.

☑ accurate   ☑ accurate   ☒ accurate
☑ precise   ☒ precise   ☒ precise

Percentage error and accuracy

The accuracy of an experiment can be analysed through the **percentage error** given by:

$$\text{percentage error} = \frac{|\text{true value} - \text{experimental value}|}{\text{true value}} \times 100\%$$

Example:
The accepted value for the ideal gas constant is 8.314 J mol^{-1} K^{-1}.

Experiment 1	Experiment 2
8.317 J mol^{-1} K^{-1}	8.103 J mol^{-1} K^{-1}
percentage error $= \left\|\dfrac{8.317 - 8.314}{8.314}\right\| \times 100$ $= 0.03608 = \mathbf{0.04}$	percentage error $= \left\|\dfrac{8.103 - 8.314}{8.314}\right\| \times 100$ $= 2.53788 = \mathbf{2.54}$

Experiment 1 has a more **accurate** result.

Uncertainty in measurement

Laboratory apparatus used for quantitative analysis usually have nominal random uncertainties which reflect the tolerances used in their manufacture. When performing experiments, always note down the uncertainty of specific pieces of equipment. The typical uncertainties of common apparatus are as follows:

Apparatus	Tolerance	
	A-grade	B-grade
Volumetric flask 250.00 cm^3	± 0.15 cm^3	± 0.30 cm^3
Volumetric flask 100.00 cm^3	± 0.10 cm^3	± 0.15 cm^3
Volumetric flask 25.00 cm^3	± 0.04 cm^3	± 0.06 cm^3
Pipette 25.00 cm^3		± 0.06 cm^3
Pipette 10.00 cm^3		± 0.04 cm^3
Burette 50.00 cm^3 with 0.10 cm^3 graduations	± 0.05 cm^3	± 0.10 cm^3

When the uncertainty is not quoted on the apparatus, the following general rules apply:

Single analogue measurements
The random error is **half the value of the smallest graduation** e.g. the reading off a burette with 0.1 cm^3 graduations should be quoted as 43.60 ± 0.05 cm^3.

Single digital measurements
The random error is the **value of the smallest graduation** or the **manufacturer's specification** e.g. the reading off a top pan balance which gives a reading up to 2 decimal places should be quoted as 45.22 ± 0.01 g.

Comparison between accuracy and precision

Precision is a **measure of the certainty** of the value determined and usually quoted as a ± **value**.

Example:
The accepted value for the ideal gas constant is 8.314 J mol^{-1} K^{-1}.

Experiment 1	Experiment 2
8.34 ± 0.03 J mol^{-1} K^{-1}	8.513 ± 0.006 J mol^{-1} K^{-1}

Experiment 1 has a more **accurate** result (closer to the true value).
Experiment 2 has a more **precise** result (has a smaller uncertainty).

Absolute, relative and percentage relative uncertainty

When an apparatus measures the following value, A ± ΔA, ΔA is known as the **absolute uncertainty**.
The **relative uncertainty** is given by:

$$\text{relative uncertainty} = \frac{\Delta A}{A}$$

The **percentage relative uncertainty** is given by:

$$\text{percentage relative uncertainty} = \frac{\Delta A}{A} \times 100\%$$

Example: For 28.5 ± 0.05,

$$\text{percentage relative uncertainty} = \left|\frac{0.05}{28.5}\right| \times 100 = \pm 0.18\%$$

Propagation of uncertainties

For the following measurements and absolute uncertainties:
$A \pm \Delta A$, $B \pm \Delta B$, $C \pm \Delta C$, etc.

Addition and Subtraction:

$$\text{propagated uncertainty} = \sqrt{\Delta A^2 + \Delta B^2 + \Delta C^2 + \ldots}$$

Multiplication and division:

$$\text{propagated \% relative uncertainty} = \sqrt{\left(\frac{\Delta A}{A} \times 100\right)^2 + \left(\frac{\Delta B}{B} \times 100\right)^2 + \left(\frac{\Delta C}{C} \times 100\right)^2 + \ldots}$$

Raising A to a power of n:

$$\text{propagated uncertainty} = \sqrt{n\left(\frac{\Delta A}{A}\right)^2}$$

Example 1
Initial burette reading = 0.65 ± 0.05 cm³
Final burette reading = 16.30 ± 0.05 cm³

relative uncertainty in volume = $\sqrt{(0.05)^2 + (0.05)^2} = 0.07$

Volume = (15.65 ± 0.07) cm³

Example 2
X = A (B – C)
A = 12.3 ± 0.5; B = 12.7 ± 0.2; C = 4.3 ± 0.1

	A	B – C	X
Value	12.3	12.7 - 4.3 = **8.4**	123 x (12.7 - 4.3) = **1033.2**
Uncertainty	0.5	$\sqrt{0.2^2 + 0.1^2} = \mathbf{0.2}$	?

percentage relative uncertainty in X = $\sqrt{\left(\dfrac{0.5}{12.3} \times 100\right)^2 + \left(\dfrac{0.2}{0.4} \times 100\right)^2} = 4.7\%$

$$\Delta X = 1033.2 \times \frac{4.7}{100} = 48.7$$

Therefore, X = 1033.2 ± 48.7
- The value is then rounded off to a similar number of decimal places.

So **X = 1033 ± 49**

Example 3
2.59 g $NiCl_2$ is dissolved in 100.00 cm³ of water. What is the concentration of this solution?

$\Delta m(NiCl_2) = \sqrt{(0.01)^2 + (0.01)^2} = 0.01$ g

% relative uncertainty in $m(NiCl_2) = \dfrac{0.01}{2.59} \times 100 = 0.39\%$

$n(NiCl_2) = \dfrac{2.59}{58.69 + (2 \times 35.45)} = 0.01999 \approx 0.0200$ mol

$\Delta n(NiCl_2) = \dfrac{0.39}{100} \times 0.0200 = 0.0001$ mol

$\therefore n(NiCl_2) = (0.0200 \pm 0.0001)$ mol

% relative uncertainty in $V(H_2O) = \dfrac{0.05}{100.00} \times 100 = 0.05\%$

$[NiCl_2] = \dfrac{0.020}{100.00 \times 10^{-3}} = 0.200$ mol dm⁻³

% relative uncertainty in $[NiCl_2] = \sqrt{0.39^2 + 0.05^2} = 0.39\%$

$\Delta[NiCl_2] = \dfrac{0.200 \times 0.63}{100} = 0.001$ mol dm⁻³

$\therefore [NiCl_2] = (0.200 \pm 0.001)$ **mol dm⁻³**

Example 4
Q=mcΔT
Initial temperature = 21.6 ± 0.1 °C
Final temperature = 24.2 ± 0.1 °C
m = 200.0 ± 0.5 g
c = 4.183 ± 0.005 J g⁻¹ K⁻¹
Amount of limiting reagent, n = 0.0500 ± 0.0005 mol

$\Delta(\Delta T) = \sqrt{(0.1)^2 + (0.1)^2} = 0.1°C \qquad \therefore \Delta T = (2.6 \pm 0.1)°C$

$Q = 200.0 \times 4.183 \times 2.6 = 21751.6$ J

% relative uncertainty in $\Delta Q = \sqrt{\left[\left(\dfrac{0.5}{200.0} \times 100\right)^2 + \left(\dfrac{0.005}{4.183} \times 100\right)^2 + \left(\dfrac{0.1}{2.6} \times 100\right)^2\right]} = 3.9\%$

% relative uncertainty in $\Delta H = \sqrt{\left[3.9^2 + \left(\dfrac{0.0005}{0.0500} \times 100\right)^2\right]} = 4.0\%$

$\Delta H = \dfrac{21751.6}{0.0500} = 435032$ J mol⁻¹ $\qquad \Delta(\Delta H) = \left|-435032 \times \dfrac{4.0}{100}\right| = 17401$ J mol⁻¹

$\Delta H = (435 \pm 17)$ **kJ mol⁻¹**

Suggested improvements

In Examples 1-3, improvements should be suggested for the methods related to all the quantities measured.
In Example 4, the main area for improvement is in the measurement of ΔT. This could include:
using a more precise thermometer e.g. up to ± 0.01 °C.
increasing the temperature range e.g. using more concentrated reagents. There is little point in trying to reduce the uncertainties in the other values until the uncertainty in the temperature has been significantly reduced.

Discrepancy from literature (accepted) value

Suppose that in Example 2, the literature value for ΔH is -562 kJ mol^{-1}. The percentage error would be $(56.2-43)/56.2 \times 100 = 23.5\%$. This value is much greater than could be accounted for by the uncertainties in the experimental value (7.7%). Suggestions must be made to account for systematic errors. These could include:
heat loss to surroundings – cover the calorimeter or use a Styrofoam calorimeter.
not taking into consideration the heat capacity of the calorimeter – measure the heat capacity of the calorimeter.

11.2 Graphical techniques
Syllabus
Nature of science – can you relate this topic to these concepts?
The idea of correlation – can be tested in experiments whose results can be displayed graphically.
Understandings – how well can you explain these statements?
Graphical techniques are an effective means of communicating the effect of an independent variable on a dependent variable, and can lead to determination of physical quantities.
Sketched graphs have labelled but unscaled axes, and are used to show qualitative trends, such as variables that are proportional or inversely proportional.
Drawn graphs have labelled and scaled axes, and are used in quantitative measurements.
Applications and skills – how well can you do all of the following?
Drawing graphs of experimental results including the correct choice of axes and scale.
Interpretation of graphs in terms of the relationships of dependent and independent variables.
Production and interpretation of best-fit lines or curves through data points, including an assessment of when it can and cannot be considered as a linear function.
Calculation of quantities from graphs by measuring slope (gradient) and intercept, including appropriate units.

Good graphs have these characteristics:
- A meaningful title.
- Labels (quantities and units) on the axes.
    - The independent variable is usually on the horizontal axis.
    - The dependent variable is usually on the vertical axis.
- Uses at least 2/3 of the space available, including extrapolation.
- Sensible, linear scales e.g. 1 unit to 2, 4, 5, 8 or 10 squares.
- A line of best fit (straight or curve) passing as near to as many of the points as possible. It does not have to go through any points.

Points to consider:
- Does the graph go through the origin? (a frequent error)
- Which points are clearly anomalous?
- Are there enough points to show a clear trend?

Linear graphs

Care must be taken when determining the relationship between the variables. The trend is usually made more obvious by including more points in the graph.

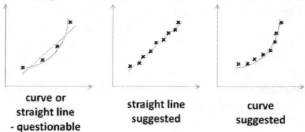

curve or straight line - questionable

straight line suggested

curve suggested

Straight line graphs have the equation **y = mx+c** can be used to determine the:
o   gradient, m
o   intercept on the vertical axis, c

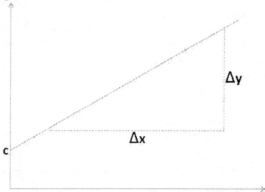

gradient, m = Δy/Δx

Extrapolation of linear graphs
Sometimes, a line has to be extended beyond the range of measurements of the graph – it has to be extrapolated.
Example: a volume-temperature graph for an ideal gas can be extrapolated backwards to determine absolute zero.

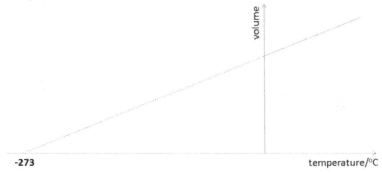

Interpolation on curves

A trend line is assumed to be present between two points of a curve. The gradient of the curve at any point is the gradient of the tangent to the curve at that point.

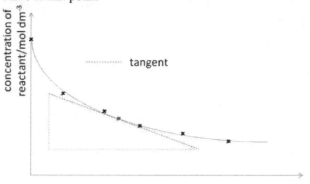

Example 1 – Boyle's law for ideal gases

$$PV = nRT$$

$$P = nRT\left(\frac{1}{V}\right)$$

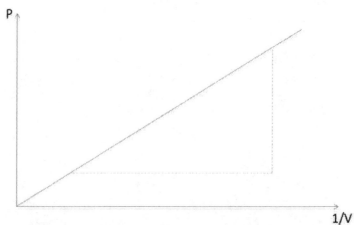

A graph of P against 1/V at constant temperature would give a straight line with gradient = nRT.

Example 2 – Charles's law for ideal gases

$$PV = nRT$$

$$V = \left(\frac{nR}{P}\right)T$$

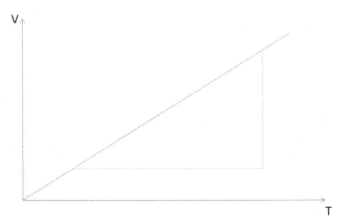

A graph of V against T at constant pressure would give a straight line with gradient = nR/P.

Example 3 – Gay-Lussac's law for ideal gases

$$PV = nRT$$
$$P = \left(\frac{nR}{V}\right)T$$

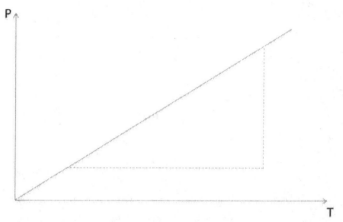

A graph of P against T at constant pressure would give a straight line with gradient = nR/V. This is not practicable as it is very difficult to measure pressure accurately.

Example 4 – Cooling curves
The portion of the graph following the maximum temperature is extrapolated backwards.

This estimates the temperature rise that would have occurred had the reaction had been instantaneous and involved no heat loss.

Example 5 – Reaction rate graphs
These are graphs of concentration (or some other property proportional to it) against time.

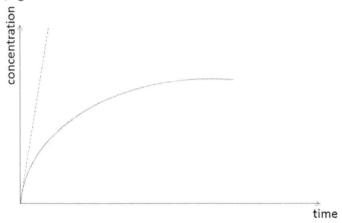

The gradient of the tangent at t=0 gives the initial rate. The gradient of any other line interpolated on the rest of the curve gives the rate of reaction at that time.

Example 6 – Activation energy graphs

The activation energy (E_a) and temperature is related by the Arrhenius equation.

$$k = Ae^{-\frac{E_a}{RT}}$$

k = rate constant
A = pre-exponential factor
E_a = activation energy
R = gas constant
T = temperature in K

Taking natural logarithms and rearranging:

$$ln\,k = ln\,A - \left(\frac{E_a}{R}\right)\left(\frac{1}{T}\right)$$

A plot of ln k against 1/T gives a straight line.

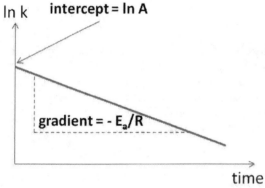

$A = e^{intercept}$
E_a = - gradient x R

11.3 Spectroscopic identification of organic compounds
Syllabus
Nature of science – can you relate this topic to these concepts?
Improvements in instrumentation – mass spectrometry, proton nuclear magnetic resonance and infrared spectroscopy have made identification and structural determination of compounds routine.
Models are developed to explain certain phenomena that may not be observable – for example, spectra are based on the bond vibration model.
Understandings – how well can you explain these statements?
The degree of unsaturation or index of hydrogen deficiency (IHD) can be used to determine from a molecular formula the number of rings or multiple bonds in a molecule.
Mass spectrometry (MS), proton nuclear magnetic resonance spectroscopy (^{1}H NMR) and infrared spectroscopy (IR) are techniques that can be used to help identify compounds and to determine their structure.
Applications and skills – how well can you do all of the following?
Determination of the IHD from a molecular formula.
Deduction of information about the structural features of a compound from percentage composition data, MS, ^{1}H NMR or IR.

Index of Hydrogen Deficiency (IHD)

Index of Hydrogen Deficiency is also known as "degree of unsaturation". It is an indication of how many molecules of H_2 need to be added to a molecule to form a saturated, non-cyclic species i.e. each H_2 molecule added counts as one degree of saturation. For molecules which are neutral, the IHD is always an integer.
The IHD can be determined from:
a molecular formula
a drawn structure
For the IB, you will need to use the method using the molecular formula.
Consider the following generic molecular formula $C_cH_hN_nO_oX_x$, where X is a halogen atom.
$IHD = 0.5 \times (2c + 2 - h - x + n)$

Mass spectroscopy

As seen in Topic 2, mass spectroscopy can be used to determine:
- the **relative atomic mass** of elements from the **relative abundances** of their isotopes.
- the **relative formula mass** of molecules from the **molecular ion peak, M⁺**.

Another use of mass spectroscopy is the analysis of the **fragmentation pattern**. During ionisation to form the molecular ion, the sample molecule can also be broken up into smaller fragments - fragmentation always results in an ion and a free radical. The peaks that appear in the spectra correspond to the following fragments:

Mass lost	Fragment lost
15	CH_3
17	OH
18	H_2O
28	$CH_2=CH_2$, $C=O$
29	CH_3CH_2, CHO
31	CH_3O
45	$COOH$

Propanal: $C_2H_5CHO^+ \rightarrow C_2H_5^+ + CHO\bullet$ or $C_2H_5CHO^+ \rightarrow C_2H_5\bullet + CHO^+$

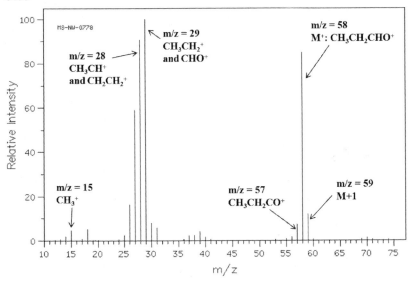

Propanone: $CH_3COCH_3^+ \rightarrow CH_3^+ + CH_3CO\bullet$ or $CH_3COCH_3^+ \rightarrow CH_3\bullet + CH_3CO^+$

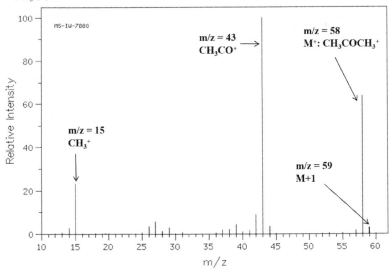

Infrared (IR) spectroscopy

Electromagnetic (em) radiation consists of an oscillating electric and magnetic field with a wide range of frequencies.
All electromagnetic waves travel at the same speed, which is $c = 3 \times 10^8$ ms^{-1}.
Different types of waves have different wavelengths (λ) and frequencies (υ).
The frequency of a wave is the number of waves which pass a particular point in 1 s.
υ is inversely proportional to λ and related by:
$c = \upsilon\lambda$
The colour and energy of the light is related to the energy of its photon by:
$E = h\upsilon$
h = Planck's constant = 6.63×10^{-34} J s
The frequency can also be expressed as a wavenumber:
wavenumber (cm^{-1}) = $1/\lambda$
(λ in cm)
Infrared radiation has a lower frequency than visible light. It is more convenient to use wavenumber for radiation in this range of frequencies.
$c = \upsilon\lambda \Rightarrow \lambda = c/\upsilon$
Therefore wavenumber = f/c
The infrared radiation used in spectroscopy typically has a range of 4000 – 400 cm^{-1}.

Infrared (IR) spectroscopy – the spring model and vibrational energy

A molecule can absorb energy in a number of ways:

Absorption mode	Energy source	Effect on atom or molecule
Translational energy	Heat	a molecule gains kinetic energy and moves faster
Rotational energy	Microwaves	a molecule rotates faster about its centre of mass
Vibrational energy	Infrared radiation	the atoms in a molecule vibrates faster or more violently
Electron excitement	Visible or ultraviolet radiation	an electron in an atom or bond is excited to a higher energy level

Polar bonds can absorb infrared radiation. Oxygen and nitrogen do not absorb infrared radiation and are not greenhouse gases whereas carbon dioxide and methane are.

Infrared (IR) spectroscopy – spectrum analysis

Infrared spectroscopy is a useful tool for discovering which functional groups are present in a compound. Many spectrometers have a large database of known compounds installed in memory and can often identify what a substance is.

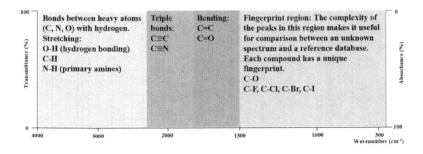

1-chloropropane, $CH_3CH_2CH_2Cl$

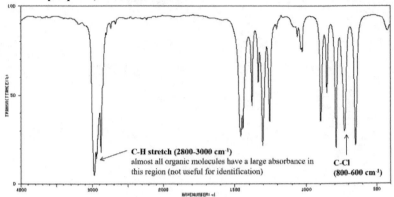

The C-Cl absorption peaks are in the fingerprint region and cannot be confirmed unless compared with a known database of chloroalkanes.

ethanol, C_2H_5OH

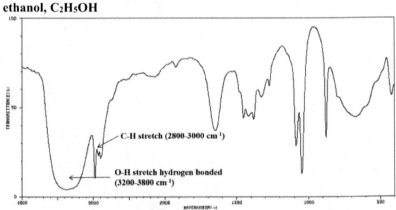

The O-H absorption peak is broad – it narrows and moves to about 3400 cm^{-1} if ethanol is dissolved in a solvent (non hydrogen bonded).

propan-2-ol (CH₃CH(OH)CH₃)

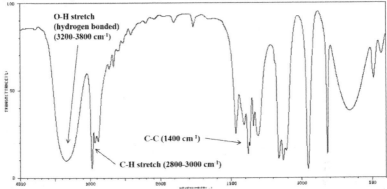

The absorption peaks at around 1400 cm⁻¹ are due to C-C bonds which are present in almost all organic molecules. Like the C-H peak, it is of limited use.

ethanoic acid, CH₃COOH

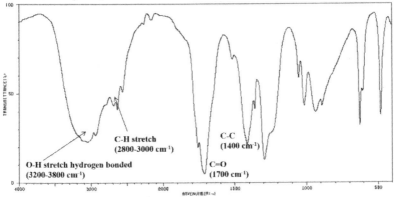

The O-H peak is due to hydrogen bonding and the formation of a dimer.

$$R - C\begin{matrix}O \cdots H - O\\ \\O - H \cdots O\end{matrix}C - R$$

ethanamine, $CH_3CH_2NH_2$

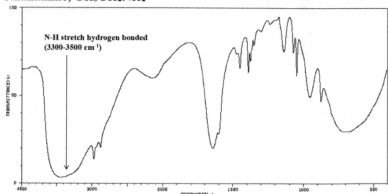

propanal, CH_3CH_2CHO

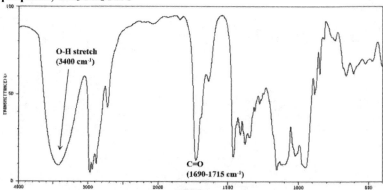

The O-H stretch is due to the equilibrium in the liquid:

$CH_3CH_2CHO \rightleftharpoons CH_3CH=CHOH$

This is not present in the spectrum for propanone.

propanone, CH₃COCH₃

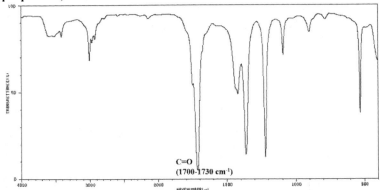

methylbenzene, C₆H₅CH₃

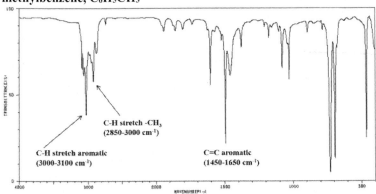

Infrared (IR) spectroscopy – progress of reaction

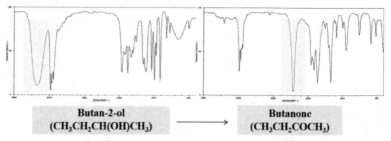

The progress of a reaction can be confirmed by comparing the IR spectra of the reactants and products.

In the reaction above, the formation of butanone from butan-2-ol can be confirmed by:
- the presence of the C=O absorption peak at 1700-1730 cm^{-1}.
- the absence of the broad O-H stretch absorption at 3200-3800 cm^{-1}.

Preparation of ibuprofen

$$H_3C-CH-CH_2-\bigcirc-CH-CH_2OH + 2[O]$$
$$\quad\quad CH_3 \quad\quad\quad\quad CH_3$$

$$\longrightarrow H_3C-CH-CH_2-\bigcirc-CH-COOH + H_2O$$
$$\quad\quad\quad CH_3 \quad\quad\quad\quad CH_3$$

The peak at 1720 cm^{-1} due to the C=O bond is present in ibuprofen but not in the starting primary alcohol.

Preparation of butanenitrile

$$CH_3CH_2CH_2Cl + KCN \rightarrow CH_3CH_2CH_2C\equiv N + KCl$$
1-chloropropane   butanenitrile

The peak at 2260 cm^{-1} due to the C≡N bond is present in butanenitrile but not in 1-chloropropane.

¹H NMR spectroscopy

The nuclei of hydrogen atoms are said to be spinning and a spinning charge generates a magnetic field. There are two possible spin states of the hydrogen nuclei. When an external magnetic field is applied, the two spin states diverge in energy into:
- parallel alignment (+½) which is of lower energy.
- anti-parallel alignment (-½) which is of higher energy.

The difference in energy between the two spin states is ΔE. The magnitude of ΔE depends:
- the strength of the applied magnetic field.
- the chemical environment of the hydrogen atom within the molecule.

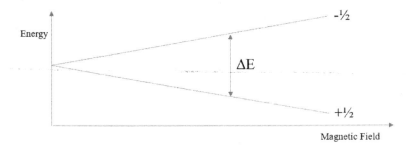

Each peak in a ¹H NMR spectrum corresponds to a unique hydrogen environment and can give clues to the following:
- the position or functional group that the hydrogen belongs to.
- the relative number of hydrogen atoms within the same environment.

The position of the peak is known as the **chemical shift, δ** and is shown in parts per million (ppm). They are all relative to the chemical shift of tetramethylsilane, TMS which has been designated as 0 ppm. TMS has 12 protons which are all equivalent and four carbons, which are also all equivalent. This means that it gives a single, strong signal in the spectrum, which turns out to be outside the range of most other signals, especially from organic compounds.

C₄H₈O₂

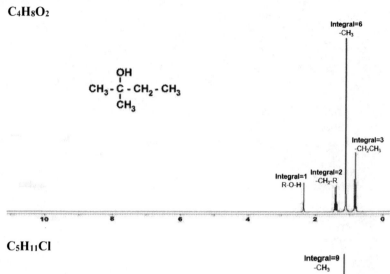

C₅H₁₁Cl

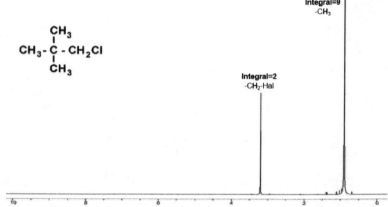

C₅H₁₀O₂

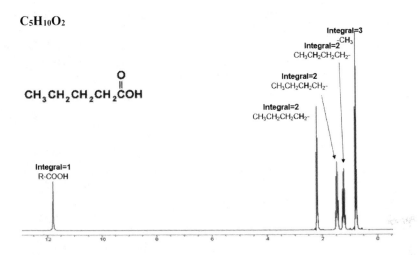

C₈H₁₂O₄

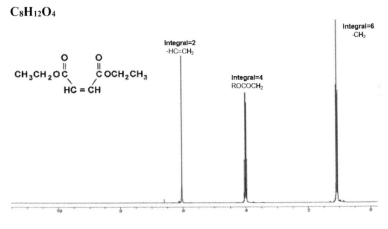

21.1 Spectroscopic identification of organic compounds
Syllabus
Nature of science – can you relate this topic to these concepts?
Improvements in modern instrumentation—advances in spectroscopic techniques (IR, ^{1}H NMR and MS) have resulted in detailed knowledge of the structure of compound.

Understandings – how well can you explain these statements?
Structural identification of compounds involves several different analytical techniques including IR, ^{1}H NMR and MS.

In a high resolution ^{1}H NMR spectrum, single peaks present in low resolution can split into further clusters of peaks.

The structural technique of single crystal X-ray crystallography can be used to identify the bond lengths and bond angles of crystalline compounds.

Applications and skills – how well can you do all of the following?
Explanation of the use of tetramethylsilane (TMS) as the reference standard.

Deduction of the structure of a compound given information from a range of analytical characterization techniques (X-ray crystallography, IR, ^{1}H NMR and MS).

High resolution ¹H NMR spectroscopy – spin-spin coupling

In high resolution ¹H NMR spectroscopy, the NMR active protons in different environments interact with each other magnetically. This is because each proton generates its own magnetic field which can adopt one of these alignments with the applied magnetic field, B_o:
- 50% of the nuclei will be aligned with B_o.
- 50% of the nuclei will be opposed to B_o.

These alignments then interact with the protons on the neighbouring carbon atom. This causes the peaks in the spectrum to split and shift upfield (lower chemical shift) or downfield (higher chemical shift).

Consider the following molecule:

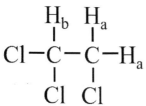

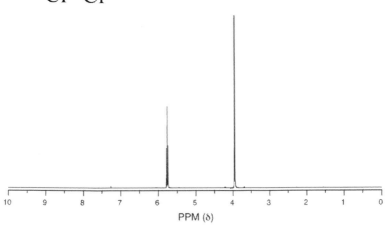

High resolution ¹H NMR spectroscopy – sample analysis

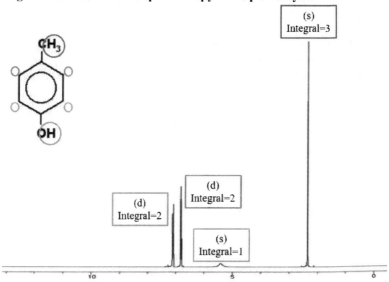

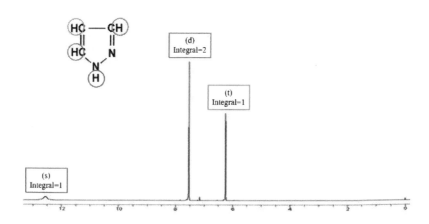

High resolution ¹H NMR spectroscopy – TMS as the solvent

The position of the peak is known as the **chemical shift, δ** and is shown in parts per million (ppm). They are all relative to the chemical shift of tetramethylsilane, TMS which has been designated as 0 ppm. It is a usual choice as a reference because:
- TMS has 12 protons which are all equivalent and four carbons, which are also all equivalent. This means that it gives a single, strong signal in the spectrum.
- The signal for TMS appears upfield and is outside the range of most other signals, especially from organic compounds.
- TMS is inert and will not interfere or react with the sample.
- TMS can be isolated from the sample after analysis as it has a very low boiling point.

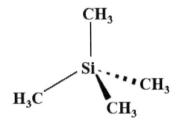

X-ray crystallography

X-ray crystallography is a technique where beams of X-ray are targeted at a crystal, which causes the beams to diffract into many specific directions.
The relative angles and intensities of the diffracted beams can be used to generate a 3-D model of the atomic and molecular structure of the crystal. The model can then be used to determine:
- **bond angles** within the crystal.
- **bond lengths** within the crystal.
- **shape** of the crystalline compound.

Made in the USA
San Bernardino, CA
04 December 2018